**William D. Clark, Ph.D.,**
**and**
**Sandra Luna McCune, Ph.D.**

New York Chicago San Francisco Lisbon London Madrid Mexico City
Milan New Delhi San Juan Seoul Singapore Sydney Toronto

The McGraw-Hill Companies



1 2 3 4 5 6 7 8 9 10 11 12 13 14 15 WDQ/WDQ 1 9 8 7 6 5 4 3 2 1 0

ISBN 978-0-07-163815-9
MHID 0-07-163815-6

Library of Congress Control Number: 2009941936



Interior design by Village Bookworks, Inc.
Interior illustrations by Glyph International



This book is printed on acid-free paper.

*To Shirley and Donice*
*You are forever in our hearts.*

# Contents

# Preface

*Practice Makes Perfect: Calculus* is designed as a tool for review and practice in calculus for the advanced beginner or intermediate learner of calculus. It is not intended to introduce concepts, but rather it is meant to reinforce what already has been presented to readers. To that end, it is a useful supplementary text for introductory courses in calculus. It can also serve as a refresher text for readers who need to revitalize previously acquired calculus skills.

Like most topics worth knowing, learning calculus requires diligence and hard work. The foremost purpose of *Practice Makes Perfect: Calculus* is as a source of solved calculus problems. We believe that the best way to develop accuracy and speed in calculus is to work numerous practice exercises. This book has more than 500 practice exercises from beginning to end. A variety of exercises and levels of difficulty are presented to provide reinforcement of calculus concepts. In each unit, a concept discussion followed by example problems precedes each set of exercises to serve as a concise review for readers already familiar with the topics covered. Concepts are broken into basic components to provide ample practice of fundamental skills.

To use *Practice Makes Perfect: Calculus* in the most effective way, it is important that you work through every exercise. After working a set of exercises, use the worked-out solutions to check your understanding of the concepts. We sincerely hope this book will help you acquire greater competence and confidence in using calculus in your future endeavors.

# LIMITS

The fundamental idea of the calculus is the concept of limit. The exercises in Part I are designed to improve your understanding and skills in working with this concept. The symbolisms involved are useful contractions/abbreviations and recognizing the "form" of these is essential in successfully producing required results. Before you begin, if you need a review of functions, see Appendix A: Basic functions and their graphs.

# The limit concept

## Limit definition and intuition

A function $f(x)$ is said to have a **limit** $A$ as $x$ approaches $c$ written $\lim_{x \to c} f(x) = A$, provided the error between $f(x)$ and $A$, written $|f(x) - A|$, can be made less than any preassigned positive number $\varepsilon$ whenever $x$ is close to, but not equal to, $c$. Heuristically, "The limit of $f$ at the point $c$ is $A$ if the value of $f$ gets near $A$ when $x$ is near $c$." We will explore this definition intuitively through the following examples.

Compute the value of $f(x) = x^2 + 5$ for the following values of $x$ that are close to, but not equal to 2 in value; and then make an observation about the results.

a. $x = 2.07$ $\quad f(x) = 9.2849$
b. $x = 1.98$ $\quad f(x) = 8.9204$
c. $x = 2.0006$ $\quad f(x) = 9.00240036$

Observation: It appears that when $x$ is close to 2 in value, then $f(x)$ is close to 9 in value.

Compute the value of $f(x) = \frac{4}{x}$ for the following values of $x$ that are close to, but not equal to 0 in value, and then make an intuitive observation about the results.

a. $x = .01$ $\quad f(x) = 400$
b. $x = -.001$ $\quad f(x) = -4000$
c. $x = .001$ $\quad f(x) = 4000$

Observation: It appears that when $x$ is close to 0 in value, $f(x)$ is not close to any fixed number in value.

Using limit notation, you can represent your observation statements for the above examples, respectively, as:

$\lim_{x \to 2} x^2 + 5 = 9$ and $\lim_{x \to 0} \frac{4}{x}$ does not exist.

EXERCISE

**1·1**

*Compute the value of f(x) when x has the indicated values given in (a) and (b). For (c), make an observation based on your results in (a) and (b).*

1. $f(x) = \dfrac{x+2}{x-5}$
   a. $x = 3.001$
   b. $x = 2.99$
   c. Observation? ______________________________

2. $f(x) = \dfrac{x-5}{4x}$
   a. $x = 1.002$
   b. $x = .993$
   c. Observation? ______________________________

3. $f(x) = \dfrac{3x^2}{x}$
   a. $x = .001$
   b. $x = -.001$
   c. Observation? ______________________________

______________________________

# Properties of limits

Basic theorems that are designed to facilitate work with limits exist, and these theorems are the "bare bones" ideas you must master to successfully deal with the limit concept. Succinctly, the most useful of these theorems are the following:

If $\lim_{x\to c} f(x)$ and $\lim_{x\to c} g(x)$ both exist, then

1. The limit of the sum (or difference) is the sum (or difference) of the limits.

$$\lim_{x\to c}[f(x)+g(x)] = \lim_{x\to c} f(x) + \lim_{x\to c} g(x)$$

2. The limit of the product is the product of the limits.

$$\lim_{x\to c}[f(x)\cdot g(x)] = \lim_{x\to c} f(x)\cdot \lim_{x\to c} g(x)$$

3. The limit of a quotient is the quotient of the limits provided the denominator limit is not 0.

$$\lim_{x\to c}\frac{f(x)}{g(x)} = \frac{\lim_{x\to c} f(x)}{\lim_{x\to c} g(x)}$$

4. If $f(x) \geq 0$, then $\lim_{x\to c}\sqrt[n]{f(x)} = \sqrt[n]{\lim_{x\to c} f(x)}$ for $n > 0$
5. $\lim_{x\to c} af(x) = a\lim_{x\to c} f(x)$ where $a$ is a constant
6. $\lim_{x\to c}[f(x)]^n = \left[\lim_{x\to c} f(x)\right]^n$ for any positive integer $n$
7. $\lim_{x\to c} x = c$
8. $\lim_{x\to c}\dfrac{1}{x} = \dfrac{1}{c}$ provided $c \neq 0$

You must guard against the error of writing or thinking that $\lim_{x \to c} f(x) = f(c)$; that is, that you determine the limit by substituting $x = c$ into the expression that defines $f(x)$ and then evaluate. Recall that in the limit concept, $x$ cannot assume the value of $c$. The complete explanation requires the concept of continuity, which is discussed in Chapter 3.

PROBLEMS Evaluate the following limits.

a. $\lim_{x \to 2} \frac{3x-5}{5x+2}$

b. $\lim_{x \to 4} (3x + \sqrt{16x})$

c. $\lim_{x \to 4} \frac{x^2-16}{x-4}$

SOLUTIONS a. $\lim_{x \to 2} \frac{3x-5}{5x+2} = \frac{\lim_{x \to 2} 3x-5}{\lim_{x \to 2} 5x+2} = \frac{1}{12}$

b. $\lim_{x \to 4} (3x + \sqrt{16x}) = \lim_{x \to 4} 3x + \lim_{x \to 4} \sqrt{16x} = 3 \lim_{x \to 4} x + \sqrt{\lim_{x \to 4} 16x} = 12 + \sqrt{64} = 20$

c. $\lim_{x \to 4} \frac{x^2-16}{x-4} = \lim_{x \to 4} \frac{(x-4)(x+4)}{x-4} = \lim_{x \to 4} (x+4) = 8$

Notice that in this example, you cannot use the quotient theorem because the limit of the denominator is zero; that is, $\lim_{x \to 4} (x-4) = 0$. However, as shown, you can take an algebraic approach to determine the limit. First, you factor the numerator. Next, using the fact that for all $x \neq 4$, $\frac{(x-4)(x+4)}{x-4} = x+4$, you can simplify the fraction and then evaluate the limit. This is a useful approach that can be applied to a number of limit problems.

d. $\lim_{x \to 1} \sqrt{6x-12}$ does not exist because $6x - 12 < 0$ when $x$ is close to 1.

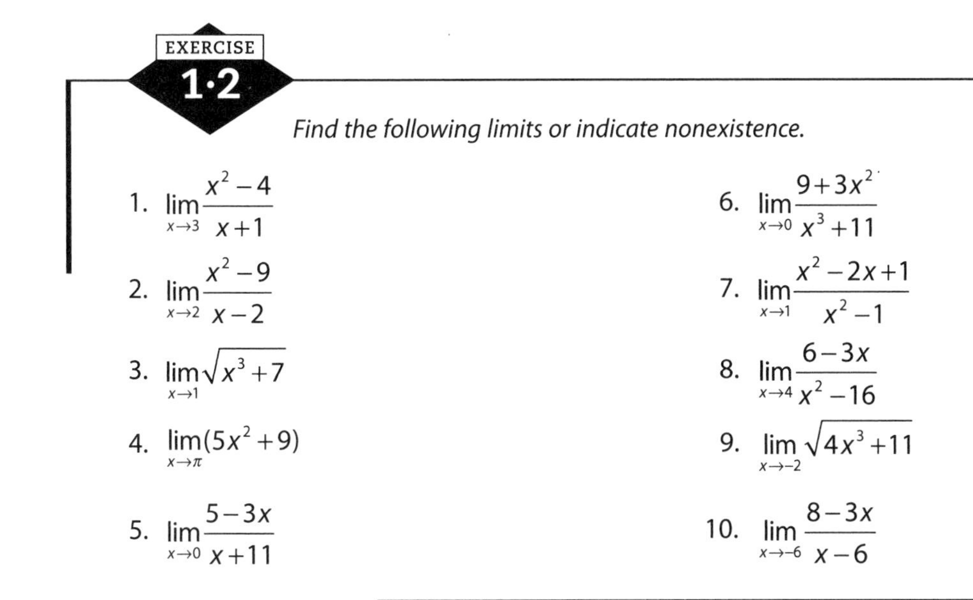

EXERCISE 1·2

*Find the following limits or indicate nonexistence.*

1. $\lim_{x \to 3} \frac{x^2-4}{x+1}$
2. $\lim_{x \to 2} \frac{x^2-9}{x-2}$
3. $\lim_{x \to 1} \sqrt{x^3+7}$
4. $\lim_{x \to \pi} (5x^2+9)$
5. $\lim_{x \to 0} \frac{5-3x}{x+11}$
6. $\lim_{x \to 0} \frac{9+3x^2}{x^3+11}$
7. $\lim_{x \to 1} \frac{x^2-2x+1}{x^2-1}$
8. $\lim_{x \to 4} \frac{6-3x}{x^2-16}$
9. $\lim_{x \to -2} \sqrt{4x^3+11}$
10. $\lim_{x \to -6} \frac{8-3x}{x-6}$

# Special limits

## Zero denominator limits

Some of the most useful limits are those in which the denominator limit is 0, even though our previous limit theorems are not directly applicable in these cases.

These types of limits can exist only if there is some sort of cancellation coming from the numerator. The key is to seek common factors of the numerator and denominator that will cancel.

PROBLEMS Evaluate the following limits.

a. $\lim_{x\to 4}\frac{x^3-8}{x-2}$

b. $\lim_{h\to 0}\frac{(5x^2+10xh+5h^2+2)-(5x^2+2)}{h}$

c. $\lim_{h\to 0}\frac{\sqrt{x+h}-\sqrt{x}}{h}$

SOLUTIONS a. $\lim_{x\to 4}\frac{x^3-8}{x-2}=\lim_{x\to 4}\frac{(x-2)(x^2+2x+4)}{x-2}=\lim_{x\to 4}(x^2+2x+4)=28$

b. $\lim_{h\to 0}\frac{(5x^2+10xh+5h^2+2)-(5x^2+2)}{h}=\lim_{h\to 0}\frac{10xh+5h^2}{h}$

$=\lim_{h\to 0}(10x+5h)=10x$

c. $\lim_{h\to 0}\frac{\sqrt{x+h}-\sqrt{x}}{h}=\lim_{h\to 0}\frac{(\sqrt{x+h}-\sqrt{x})(\sqrt{x+h}+\sqrt{x})}{h(\sqrt{x+h}+\sqrt{x})}$

$=\lim_{h\to 0}\frac{(x+h)-x}{h(\sqrt{x+h}+\sqrt{x})}=\lim_{h\to 0}\frac{h}{h(\sqrt{x+h}+\sqrt{x})}$

$=\lim_{h\to 0}\frac{1}{(\sqrt{x+h}+\sqrt{x})}=\frac{1}{2\sqrt{x}}$

PROBLEM If $f(x) = 6x^2 + 7$, then find $\lim_{h\to 0}\frac{f(x+h)-f(x)}{h}$.

SOLUTION $\lim_{h\to 0}\frac{f(x+h)-f(x)}{h}=\lim_{h\to 0}\frac{(6(x+h)^2+7)-(6x^2+7)}{h}$

$=\lim_{h\to 0}\frac{6x^2+12xh+6h^2+7-6x^2-7}{h}$

$=\lim_{h\to 0}(12x+6h)=12x$

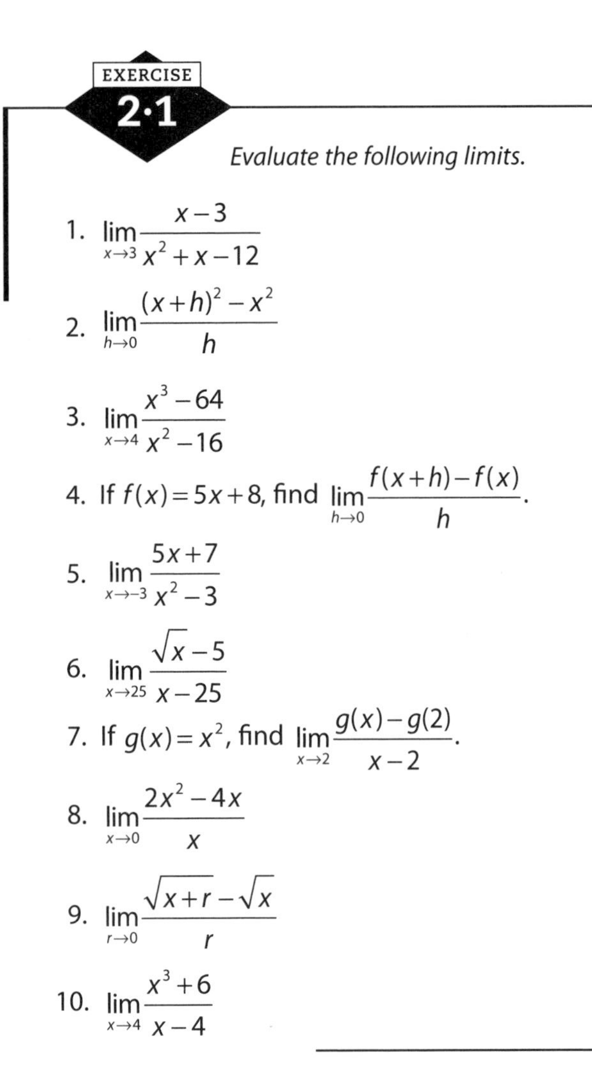

EXERCISE

**2·1**

*Evaluate the following limits.*

1. $\lim_{x\to 3} \frac{x-3}{x^2+x-12}$

2. $\lim_{h\to 0} \frac{(x+h)^2-x^2}{h}$

3. $\lim_{x\to 4} \frac{x^3-64}{x^2-16}$

4. If $f(x)=5x+8$, find $\lim_{h\to 0} \frac{f(x+h)-f(x)}{h}$.

5. $\lim_{x\to -3} \frac{5x+7}{x^2-3}$

6. $\lim_{x\to 25} \frac{\sqrt{x}-5}{x-25}$

7. If $g(x)=x^2$, find $\lim_{x\to 2} \frac{g(x)-g(2)}{x-2}$.

8. $\lim_{x\to 0} \frac{2x^2-4x}{x}$

9. $\lim_{r\to 0} \frac{\sqrt{x+r}-\sqrt{x}}{r}$

10. $\lim_{x\to 4} \frac{x^3+6}{x-4}$

# Infinite limits and limits involving infinity

The variable $x$ is said to approach $\infty$ ($-\infty$) if $x$ increases (decreases) without bound. For example, $x$ approaches $\infty$ if $x$ assumes the values 2, 3, 4, 5, 6, and so on, consecutively. Note, however, that $y$ does not approach infinity if $y$ assumes the values 2, –2, 4, –4, 6, –6, and so on, in the same manner.

A function becomes **positively infinite** as $x$ approaches $c$ if for every $M > 0$, $f(x) > M$ for every $x$ close to, but not equal to, $c$. Similarly, a function becomes **negatively infinite** as $x$ approaches $c$ if for every $M < 0$, $f(x) < M$ for every $x$ close to, but not equal to, $c$.

PROBLEMS Evaluate the following limits.

a. $\lim_{x\to\infty} \frac{a}{x}$, where $a$ is any constant

b. $\lim_{x\to\infty} x^2$

c. $\lim_{x\to -\infty} \frac{3x+12}{x-1}$

d. $\lim_{x\to 3} \frac{4}{|x-3|}$

SOLUTIONS a. $\lim_{x\to\infty}\frac{a}{x}=0$ for any constant $a$

b. $\lim_{x\to\infty}x^2=\infty$

c. $\lim_{x\to-\infty}\frac{3x+12}{x-1}=\lim_{x\to-\infty}\frac{3+\frac{12}{x}}{1-\frac{1}{x}}=\frac{3+0}{1-0}=3$

d. $\lim_{x\to3}\frac{4}{|x-3|}=\infty$

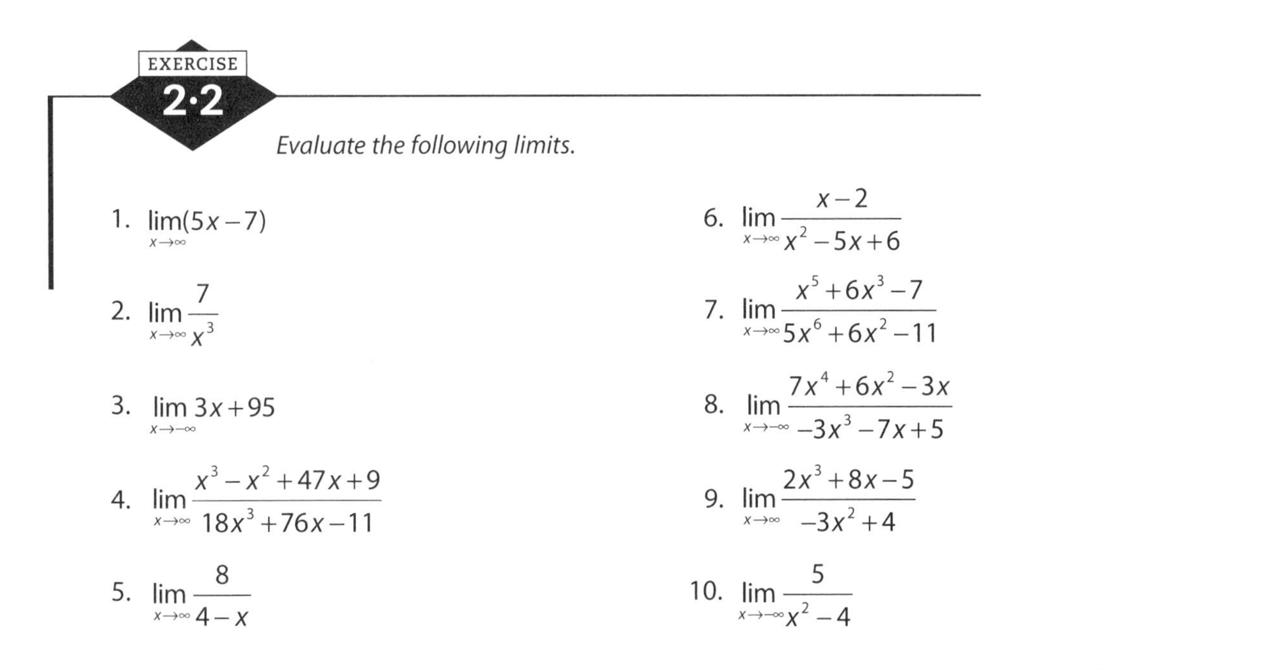

EXERCISE 2·2

*Evaluate the following limits.*

1. $\lim_{x\to\infty}(5x-7)$
2. $\lim_{x\to\infty}\frac{7}{x^3}$
3. $\lim_{x\to-\infty}3x+95$
4. $\lim_{x\to\infty}\frac{x^3-x^2+47x+9}{18x^3+76x-11}$
5. $\lim_{x\to\infty}\frac{8}{4-x}$
6. $\lim_{x\to\infty}\frac{x-2}{x^2-5x+6}$
7. $\lim_{x\to\infty}\frac{x^5+6x^3-7}{5x^6+6x^2-11}$
8. $\lim_{x\to-\infty}\frac{7x^4+6x^2-3x}{-3x^3-7x+5}$
9. $\lim_{x\to\infty}\frac{2x^3+8x-5}{-3x^2+4}$
10. $\lim_{x\to-\infty}\frac{5}{x^2-4}$

# Left-hand and right-hand limits

Directional limits are necessary in many applications and we write $\lim_{x\to c^+}f(x)$ to denote the limit concept as $x$ approaches $c$ through values of $x$ larger than $c$. This limit is called the **right-hand limit of $f$ at $c$**; and, similarly, $\lim_{x\to c^-}f(x)$ is the notation for the **left-hand limit of $f$ at $c$**.

**Theorem:** $\lim_{x\to c}f(x)=L$ if and only if $\lim_{x\to c^+}f(x)=\lim_{x\to c^-}f(x)=L$. This theorem is a very useful tool in evaluating certain limits and in determining whether a limit exists.

PROBLEMS Evaluate the following limits.

a. $\lim_{x\to3^+}\frac{4}{x-3}$

b. $\lim_{x\to1^-}\frac{15}{x-1}$

c. $\lim_{x\to2^+}[x]$ Note: $[x]$ denotes the greatest integer function (See Appendix A)

d. $\lim_{x \to 2^-} [x]$

e. $\lim_{x \to 2} [x]$

SOLUTIONS a. $\lim_{x \to 3^+} \frac{4}{x-3} = \infty$

b. $\lim_{x \to 1^-} \frac{15}{x-1} = -\infty$

c. $\lim_{x \to 2^+} [x] = 2$

d. $\lim_{x \to 2^-} [x] = 1$

e. $\lim_{x \to 2} [x]$ does not exist because $\lim_{x \to 2^+} [x] = 2$ and $\lim_{x \to 2^-} [x] = 1$, so the right- and left-hand limits are not equal.

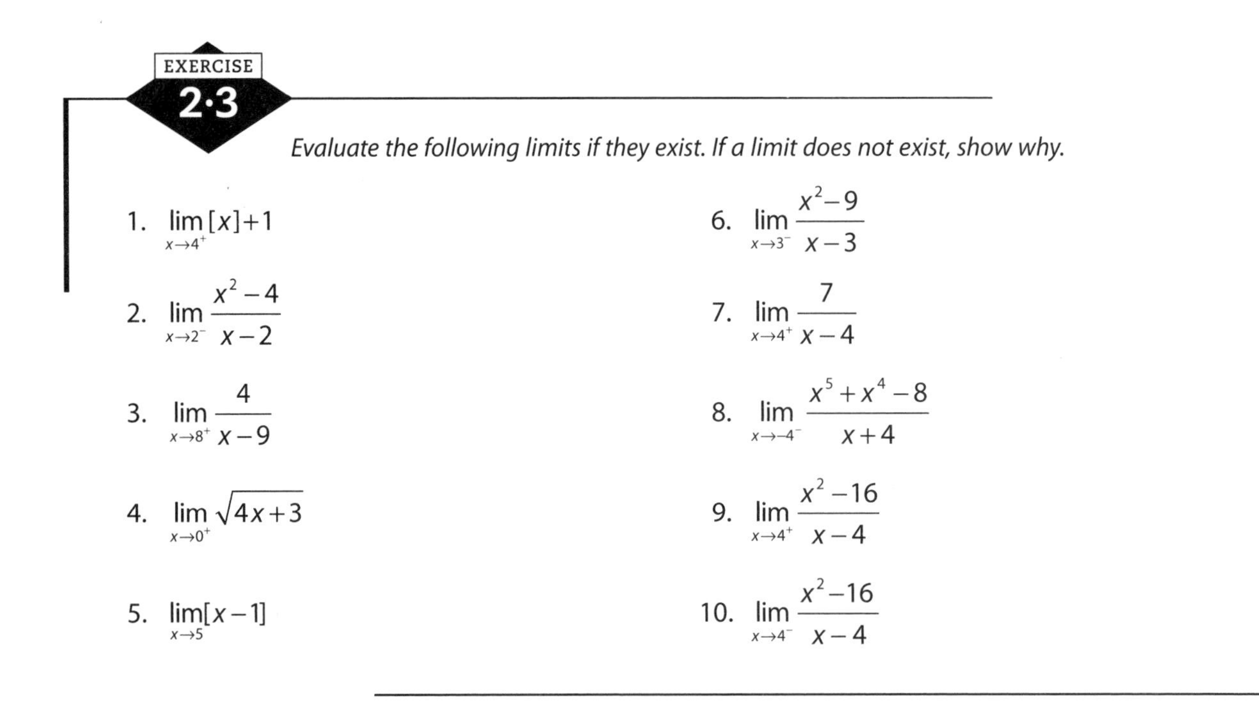

EXERCISE 2·3

*Evaluate the following limits if they exist. If a limit does not exist, show why.*

1. $\lim_{x \to 4^+} [x] + 1$
2. $\lim_{x \to 2^-} \frac{x^2 - 4}{x - 2}$
3. $\lim_{x \to 8^+} \frac{4}{x - 9}$
4. $\lim_{x \to 0^+} \sqrt{4x + 3}$
5. $\lim_{x \to 5} [x - 1]$
6. $\lim_{x \to 3^-} \frac{x^2 - 9}{x - 3}$
7. $\lim_{x \to 4^+} \frac{7}{x - 4}$
8. $\lim_{x \to -4^-} \frac{x^5 + x^4 - 8}{x + 4}$
9. $\lim_{x \to 4^+} \frac{x^2 - 16}{x - 4}$
10. $\lim_{x \to 4^-} \frac{x^2 - 16}{x - 4}$

# Continuity

## Definition of continuity

A function $f$ is **continuous** at a point $c$ if and only if

1. $f(c)$ is defined; and
2. $\lim_{x\to c} f(x)$ *exists*; and
3. $\lim_{x\to c} f(x) = f(\lim_{x\to c} x) = f(c)$.

If a function fails to satisfy any one of these conditions, then it is not continuous at $x = c$ and is said to be **discontinuous** at $x = c$.

Roughly speaking, a function is continuous if its graph can be drawn without lifting the pencil. Strictly speaking, this is not mathematically accurate, but it is an intuitive way of visualizing continuity.

Notice that when a function is continuous at a point $c$, you have the situation whereby the limit may be calculated by actually evaluating the function at the point $c$. Recall that you were cautioned against determining limits this way in an earlier discussion; however, when a function is known to be continuous at $x = c$, then $\lim_{x\to c} f(x) = f(c)$.

By its definition, continuity is a point-wise property of a function, but this idea is extended by saying that a function is **continuous on an interval** $a \le x \le b$ if and only if $f$ is continuous at each point in the interval. At the end points, the right- and left-hand limits apply, respectively, to get right and left continuity if these limits exist.

PROBLEMS Determine whether the following functions are either continuous or discontinuous at the indicated point.

a. $f(x) = \sqrt{4x+7}$ at $x = 4$

b. $f(x) = x - 4$ at $x = 3$

c. $f(x) = 3x^2 + 7$ at $x = 2$

d. $f(x) = \dfrac{12}{x-2}$ at $x = 2$

e. $f(x) = \dfrac{x^2-4}{x-2}$ at $x = 2$

SOLUTIONS a. $\lim_{x\to 4}\sqrt{4x+7}=\sqrt{\lim_{x\to 4}(4x+7)}=\sqrt{4(\lim_{x\to 4}x)+7}=\sqrt{23}$; thus, the function is continuous at 4.

b. $\lim_{x\to 3}(x-4)=((\lim_{x\to 3}x)-4)=-1$; thus, the function is continuous at 3.

c. $\lim_{x\to 2}(3x^2+7)=(3(\lim_{x\to 2}x^2)+7)=19$; thus, the function is continuous at 2.

d. $\lim_{x\to 2}\frac{12}{x-2}$ does not exist; thus, the function is discontinuous at 2.

e. $f(x)=\frac{x^2-4}{x-2}$ is discontinuous at 2 because the function is not defined at 2. However, the limit of $f(x)$ as $x$ approaches 2 is 4, so the limit exists but $\lim_{x\to 2}f(x)\neq f(2)$. If $f(2)$ is now defined to be 4 then the "new" function $f(x)=$ $\begin{Bmatrix}\frac{x^2-4}{x-2} & x\neq 2\\ 4 & x=2\end{Bmatrix}$ is continuous at 2. Since the discontinuity at 2 can be "removed," then the original function is said to have a **removable discontinuity** at 2.

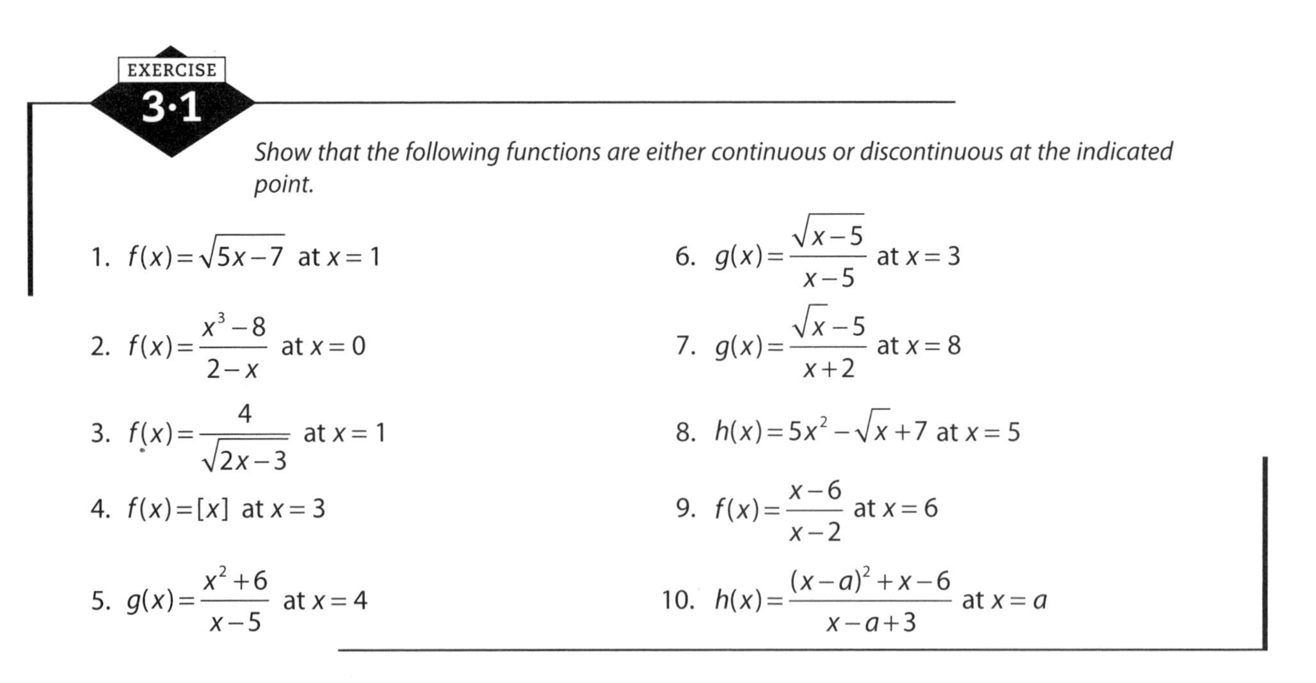

EXERCISE 3·1

*Show that the following functions are either continuous or discontinuous at the indicated point.*

1. $f(x)=\sqrt{5x-7}$ at $x=1$
2. $f(x)=\frac{x^3-8}{2-x}$ at $x=0$
3. $f(x)=\frac{4}{\sqrt{2x-3}}$ at $x=1$
4. $f(x)=[x]$ at $x=3$
5. $g(x)=\frac{x^2+6}{x-5}$ at $x=4$
6. $g(x)=\frac{\sqrt{x-5}}{x-5}$ at $x=3$
7. $g(x)=\frac{\sqrt{x}-5}{x+2}$ at $x=8$
8. $h(x)=5x^2-\sqrt{x}+7$ at $x=5$
9. $f(x)=\frac{x-6}{x-2}$ at $x=6$
10. $h(x)=\frac{(x-a)^2+x-6}{x-a+3}$ at $x=a$

# Properties of continuity

The arithmetic properties of continuity follow immediately from the limit properties in Chapter 1. If $f$ and $g$ are continuous at $x=c$, then the following functions are also continuous at $c$:

1. Sum and difference: $f\pm g$
2. Product: $fg$
3. Scalar multiple: $af$, for $a$ a real number
4. Quotient: $\frac{f}{g}$, provided $g(c)\neq 0$

Further, if $g$ is continuous at $c$ and $f$ is continuous at $g(c)$ then the composite function $f \circ g$ defined by $(f \circ g)(x) = f(g(x))$ is continuous at $c$. In limit notation, $\lim_{x \to c} f(g(x)) = f(\lim_{x \to c} g(x)) = f(g(c))$. This function composition property is one of the most important results of continuity.

If a function is continuous on the entire real line, the function is **everywhere continuous**; that is to say, its graph has no holes, jumps, or gaps in it. The following types of functions are continuous at every point in their domains:

Constant functions: $f(x) = k$, where $k$ is a constant

Power functions: $f(x) = x^n$, where $n$ is a positive integer

Polynomial functions: $f(x) = a_n x^n + a_{n-1}x^{n-1} + \cdots + a_1 x + a_0$

Rational functions: $f(x) = \dfrac{p(x)}{q(x)}$, provided $p(x)$ and $q(x)$ are polynomials and $q(x) \neq 0$

Radical functions: $f(x) = \sqrt[n]{x}, x \geq 0$, $n$ a positive integer

Trigonometric functions: $f(x) = \sin x$ and $f(x) = \cos x$ are everywhere continuous; $f(x) = \tan x$, $f(x) = \csc x$, $f(x) = \sec x$, and $f(x) = \cot x$ are continuous only wherever they exist.

Logarithm functions: $f(x) = \ln x$ and $f(x) = \log_b x, b > 0, b \neq 1$

Exponential functions: $f(x) = e^x$ and $f(x) = b^x, b > 0, b \neq 1$

PROBLEM Discuss the continuity of the following function: $g(x) = 3(\sin 3x)$ at a real number $c$.

SOLUTION $3x$ is continuous at $c$ and $\sin x$ is continuous at all real numbers and so $\sin(3x)$ is continuous at $c$ by the composition property. Finally, $3\sin(3x)$ is continuous at $c$ by the constant multiple property of continuity.

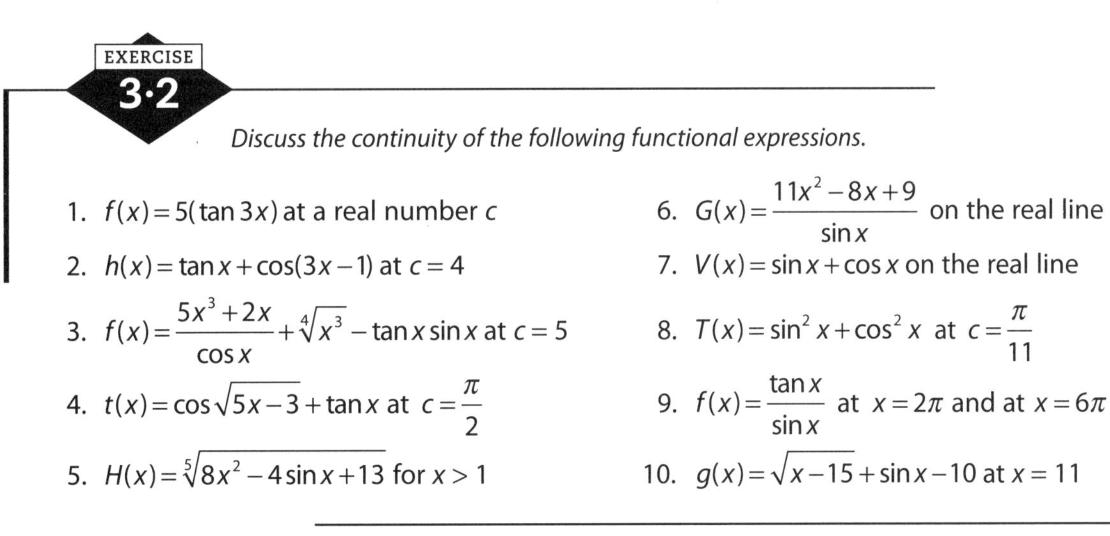

EXERCISE 3·2

*Discuss the continuity of the following functional expressions.*

1. $f(x) = 5(\tan 3x)$ at a real number $c$
2. $h(x) = \tan x + \cos(3x - 1)$ at $c = 4$
3. $f(x) = \dfrac{5x^3 + 2x}{\cos x} + \sqrt[4]{x^3} - \tan x \sin x$ at $c = 5$
4. $t(x) = \cos\sqrt{5x - 3} + \tan x$ at $c = \dfrac{\pi}{2}$
5. $H(x) = \sqrt[5]{8x^2 - 4\sin x + 13}$ for $x > 1$
6. $G(x) = \dfrac{11x^2 - 8x + 9}{\sin x}$ on the real line
7. $V(x) = \sin x + \cos x$ on the real line
8. $T(x) = \sin^2 x + \cos^2 x$ at $c = \dfrac{\pi}{11}$
9. $f(x) = \dfrac{\tan x}{\sin x}$ at $x = 2\pi$ and at $x = 6\pi$
10. $g(x) = \sqrt{x - 15} + \sin x - 10$ at $x = 11$

# Intermediate Value Theorem (IVT)

The Intermediate Value Theorem states: If $f$ is continuous on the closed interval $[a, b]$ and if $f(a) \neq f(b)$, then for every number $k$ between $f(a)$ and $f(b)$ there exists a value $x_0$ in the interval $[a, b]$ such that $f(x_0) = k$.

The Intermediate Value Theorem is a useful tool for showing the existence of zeros of a function. If a continuous function changes sign on an interval, then this theorem assures you that there must be a point in the interval at which the function takes on the value of 0. It must be noted, however, that the theorem is an existence theorem and does not locate the point at which the zero occurs. Finding that point is another problem. The following example will illustrate using the IVT to determine whether a zero exists and give some insight into finding such a point (or points).

PROBLEM Is there a number in the interval [0, 3] such that $f(x)=x^2-x-2=-1$? This question is equivalent to asking whether there is a number in [0, 3] such that $f(x)=x^2-x-1=0$.

SOLUTION The function is continuous on [0, 3], and you can see that $f(0)=0^2-0-1=-1$ and $f(3)=3^2-3-1=5$. Since $f(0)<0$ and $f(3)>0$, by the IVT, you know there must be a number in [0, 3] such that $f(x)=x^2-x-1=0$; that is, there is a solution to the problem. In this case, a solution can be found by solving the quadratic equation, $x^2-x-1=0$, to obtain the two roots: $\frac{1\pm\sqrt{5}}{2}$. Approximating these two values gives 1.62 and −0.62, of which only 1.62 is in the interval [0, 3]. Thus, there does exist a number, namely $\frac{1+\sqrt{5}}{2}$, in the interval [0, 3] such that $f\left(\frac{1+\sqrt{5}}{2}\right)=0$.

EXERCISE 3·3

*For 1–5, use the IVT to determine whether the given function has a zero in the given interval. Explain your reasoning.*

1. $f(x)=4x^4-3x^3+2x-5$ on [−2, 0]
2. $g(x)=\sqrt{9-x^2}$ on [−2.5, 2]
3. $f(x)=\frac{3}{x+4}$ on [−5, 0]
4. $g(x)=\frac{8x}{3x-5}$ on [10, 12]
5. $f(x)=\frac{x^3+x^2}{x+1}$ on [−2, 2]

*For 6–10, use the IVT to determine whether a zero exists in the given interval; and, if so, find the zero (or zeros) in the interval.*

6. $h(x)=x^2+5x-2$ on [−3, 4]
7. $g(x)=\frac{x^3-3}{x+4}$ on [0, 6]
8. $h(x)=\sin(x)$ on [−1, 1]
9. $F(x)=\cos(x)$ on [5, 8]
10. $G(x)=\frac{\sin(x)}{\cos(x)}$ on $\left[-\frac{\pi}{4},\frac{\pi}{4}\right]$

# DIFFERENTIATION

Differentiation is the process of determining the derivative of a function. Part II begins with the formal definition of the derivative of a function and shows how the definition is used to find the derivative. However, the material swiftly moves on to finding derivatives using standard formulas for differentiation of certain basic function types. Properties of derivatives, numerical derivatives, implicit differentiation, and higher-order derivatives are also presented.

# Definition of the derivative and derivatives of some simple functions

## Definition of the derivative

The **derivative** $f'$ (read "$f$ prime") of the function $f$ at the number $x$ is defined as $f'(x)=\lim_{h\to 0}\frac{f(x+h)-f(x)}{h}$, if this limit exists. If this limit does not exist, then $f$ does not have a derivative at $x$. This limit may also be written $f'(c)=\lim_{x\to c}\frac{f(x)-f(c)}{x-c}$ for the derivative at $c$.

PROBLEM Given the function $f$ defined by $f(x)=-2x+5$, use the definition of the derivative to find $f'(x)$.

SOLUTION By definition, $f'(x)=\lim_{h\to 0}\frac{f(x+h)-f(x)}{h}$

$$=\lim_{h\to 0}\frac{(-2(x+h)+5)-(-2x+5)}{h}=\lim_{h\to 0}\frac{(-2x-2h+5)+2x-5}{h}$$

$$=\lim_{h\to 0}\frac{-2x-2h+5+2x-5}{h}=\lim_{h\to 0}\frac{-2h}{h}=\lim_{h\to 0}(-2)=-2.$$

PROBLEM Given the function $f$ defined by $f(x)=x^2+2x$, use the definition of the derivative to find $f'(x)$.

SOLUTION By definition, $f'(x)=\lim_{h\to 0}\frac{f(x+h)-f(x)}{h}$

$$=\lim_{h\to 0}\frac{((x+h)^2+2(x+h))-(x^2+2x)}{h}$$

$$=\lim_{h\to 0}\frac{(x^2+2xh+h^2+2x+2h)-x^2-2x}{h}$$

$$=\lim_{h\to 0}\frac{x^2+2xh+h^2+2x+2h-x^2-2x}{h}=\lim_{h\to 0}\frac{2xh+h^2+2h}{h}$$

$$=\lim_{h\to 0}\frac{h(2x+h+2)}{h}=\lim_{h\to 0}(2x+h+2)=2x+2.$$

Various symbols are used to represent the derivative of a function $f$. If you use the notation $y = f(x)$, then the derivative of $f$ can be symbolized by $f'(x), y', D_x f(x), D_x y, \frac{dy}{dx}$, or $\frac{d}{dx}f(x)$.

**Note:** Hereafter, you should assume that any value for which a function is undefined is excluded.

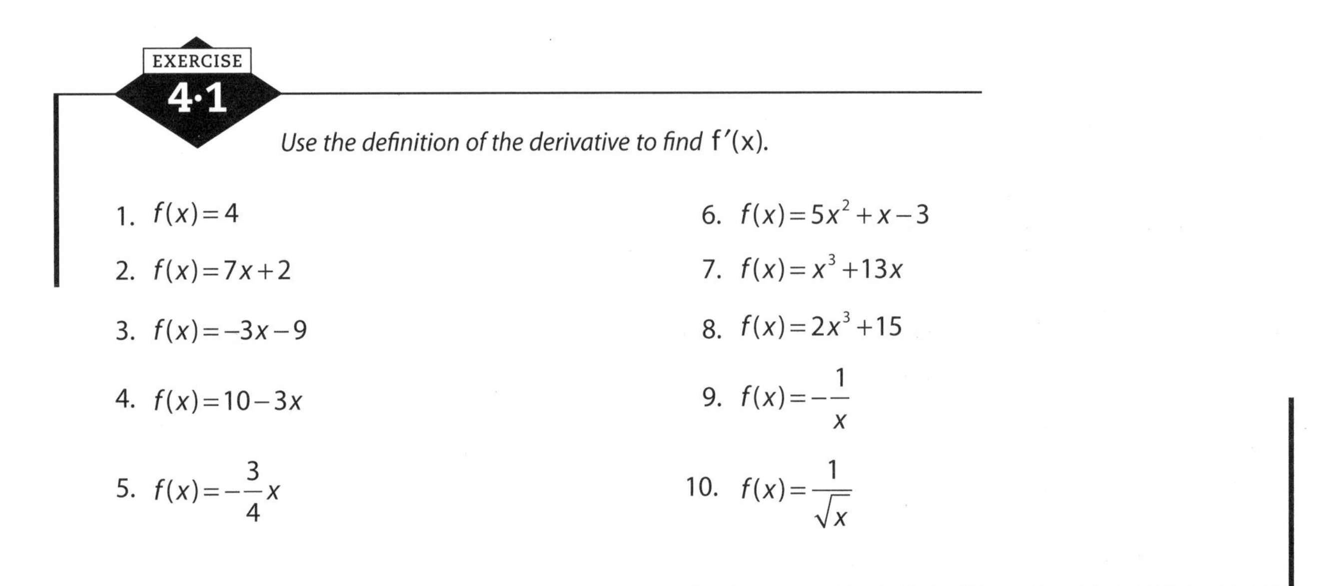

EXERCISE **4·1**

*Use the definition of the derivative to find* $f'(x)$.

1. $f(x)=4$
2. $f(x)=7x+2$
3. $f(x)=-3x-9$
4. $f(x)=10-3x$
5. $f(x)=-\frac{3}{4}x$
6. $f(x)=5x^2+x-3$
7. $f(x)=x^3+13x$
8. $f(x)=2x^3+15$
9. $f(x)=-\frac{1}{x}$
10. $f(x)=\frac{1}{\sqrt{x}}$

# Derivative of a constant function

Fortunately, you do not have to resort to finding the derivative of a function directly from the definition of a derivative. Instead, you can memorize standard formulas for differentiating certain basic functions. For instance, the derivative of a constant function is always zero. In other words, if $f(x)=c$ is a constant function, then $f'(x)=0$; that is, if $c$ is any constant, $\frac{d}{dx}(c)=0$.

The following examples illustrate the use of this formula:

- $\frac{d}{dx}(25)=0$
- $\frac{d}{dx}(-100)=0$

EXERCISE **4·2**

*Find the derivative of the given function.*

1. $f(x)=7$
2. $y=5$
3. $f(x)=0$
4. $f(t)=-3$
5. $f(x)=\pi$
6. $g(x)=25$
7. $s(t)=100$
8. $z(x)=2^3$
9. $y=-\frac{1}{2}$
10. $f(x)=\sqrt{41}$

# Derivative of a linear function

The derivative of a linear function is the slope of its graph. Thus, if $f(x) = mx + b$ is a linear function, then $f'(x) = m$; that is, $\frac{d}{dx}(mx + b) = m$.

The following examples illustrate the use of this formula:

- If $f(x) = 10x - 2$, then $f'(x) = 10$
- If $y = -2x + 5$, then $y' = -2$
- $\frac{d}{dx}\left(\frac{3}{5}x\right) = \frac{3}{5}$

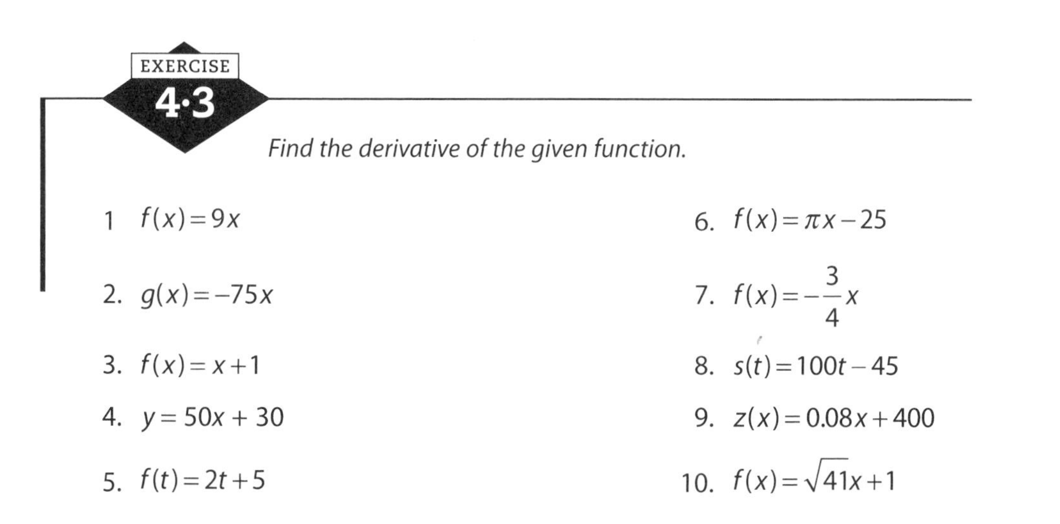

EXERCISE 4·3

*Find the derivative of the given function.*

1 $f(x) = 9x$

2. $g(x) = -75x$

3. $f(x) = x + 1$

4. $y = 50x + 30$

5. $f(t) = 2t + 5$

6. $f(x) = \pi x - 25$

7. $f(x) = -\frac{3}{4}x$

8. $s(t) = 100t - 45$

9. $z(x) = 0.08x + 400$

10. $f(x) = \sqrt{41}x + 1$

# Derivative of a power function

The function $f(x) = x^n$ is called a power function. The following formula for finding the derivative of a power function is one you will use frequently in calculus:

If $n$ is a real number, then $\frac{d}{dx}(x^n) = nx^{n-1}$.

The following examples illustrate the use of this formula:

- If $f(x) = x^2$, then $f'(x) = 2x$
- If $y = x^{\frac{1}{2}}$, then $y' = \frac{1}{2}x^{-\frac{1}{2}}$
- $\frac{d}{dx}(x^{-1}) = -1x^{-2}$

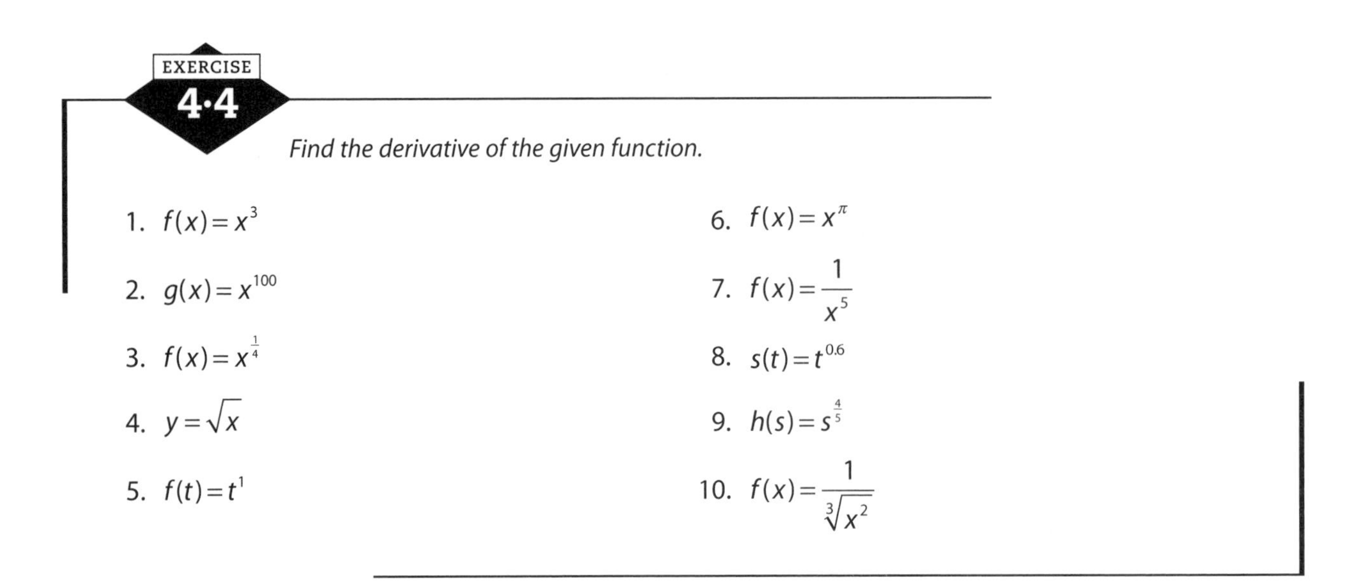

EXERCISE 4·4

*Find the derivative of the given function.*

1. $f(x)=x^3$
2. $g(x)=x^{100}$
3. $f(x)=x^{\frac{1}{4}}$
4. $y=\sqrt{x}$
5. $f(t)=t^1$
6. $f(x)=x^{\pi}$
7. $f(x)=\dfrac{1}{x^5}$
8. $s(t)=t^{0.6}$
9. $h(s)=s^{\frac{4}{5}}$
10. $f(x)=\dfrac{1}{\sqrt[3]{x^2}}$

# Numerical derivatives

In many applications derivatives need to be computed numerically. The term **numerical derivative** refers to the numerical value of the derivative of a given function at a given point, provided the function has a derivative at the given point.

Suppose $k$ is a real number and the function $f$ is differentiable at $k$, then the numerical derivative of $f$ at the point $k$ is the value of $f'(x)$ when $x=k$. To find the numerical derivative of a function at a given point, first find the derivative of the function, and then evaluate the derivative at the given point. Proper notation to represent the value of the derivative of a function $f$ at a point $k$ includes $f'(k)$, $\left.\dfrac{dy}{dx}\right|_{x=k}$, and $\left.\dfrac{dy}{dx}\right|_{k}$.

PROBLEM If $f(x)=x^2$, find $f'(5)$.

SOLUTION For $f(x)=x^2, f'(x)=2x$; thus, $f'(5)=2(5)=10$

PROBLEM If $y=x^{\frac{1}{2}}$, find $\left.\dfrac{dy}{dx}\right|_{x=9}$.

SOLUTION For $y=x^{\frac{1}{2}}, y'=\dfrac{dy}{dx}=\dfrac{1}{2}x^{-\frac{1}{2}}$; thus, $\left.\dfrac{dy}{dx}\right|_{x=9}=\dfrac{1}{2}(9)^{-\frac{1}{2}}=\dfrac{1}{2}\cdot\dfrac{1}{3}=\dfrac{1}{6}$

PROBLEM Find $\dfrac{d}{dx}(x^{-1})$ at $x=25$.

SOLUTION $\dfrac{d}{dx}(x^{-1})=-1x^{-2}$; at $x=25, -1x^{-2}=-1(25)^{-2}=-\dfrac{1}{625}$

Note the following two special situations:

1. If $f(x)=c$ is a constant function, then $f'(x)=0$, for every real number $x$; and
2. If $f(x)=mx+b$ is a linear function, then $f'(x)=m$, for every real number $x$.

Numerical derivatives of these functions are illustrated in the following examples:

- If $f(x) = 25$, then $f'(5) = 0$
- If $y = -2x + 5$, then $\left.\dfrac{dy}{dx}\right|_{x=9} = -2$

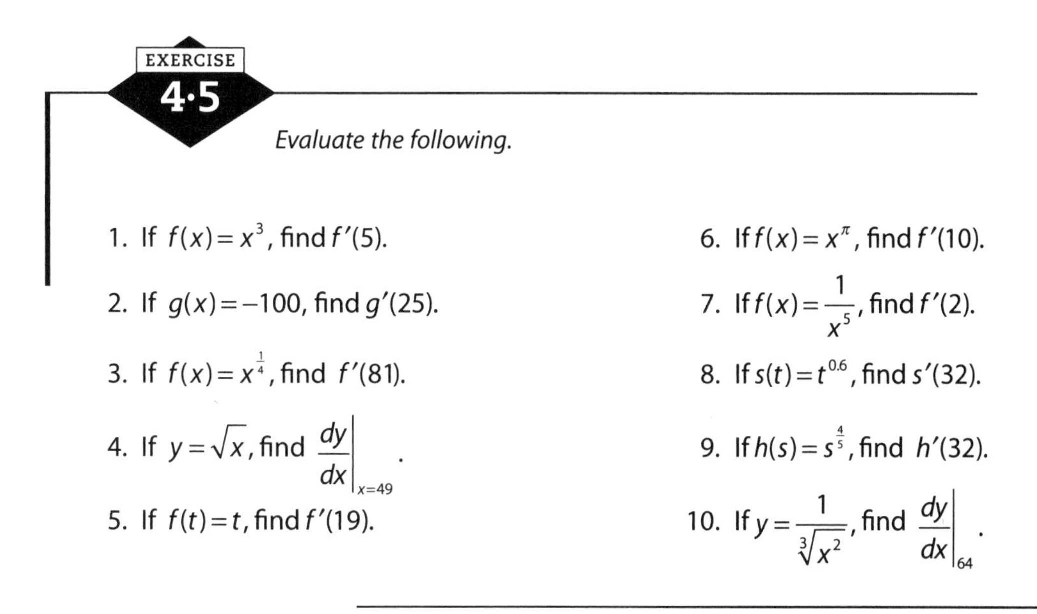

EXERCISE 4·5

*Evaluate the following.*

1. If $f(x) = x^3$, find $f'(5)$.
2. If $g(x) = -100$, find $g'(25)$.
3. If $f(x) = x^{\frac{1}{4}}$, find $f'(81)$.
4. If $y = \sqrt{x}$, find $\left.\dfrac{dy}{dx}\right|_{x=49}$.
5. If $f(t) = t$, find $f'(19)$.
6. If $f(x) = x^{\pi}$, find $f'(10)$.
7. If $f(x) = \dfrac{1}{x^5}$, find $f'(2)$.
8. If $s(t) = t^{0.6}$, find $s'(32)$.
9. If $h(s) = s^{\frac{4}{5}}$, find $h'(32)$.
10. If $y = \dfrac{1}{\sqrt[3]{x^2}}$, find $\left.\dfrac{dy}{dx}\right|_{64}$.

# Rules of differentiation

## Constant multiple of a function rule

Suppose $f$ is any differentiable function and $k$ is any real number, then $kf$ is also differentiable with its derivative given by

$$\frac{d}{dx}(kf(x)) = k\frac{d}{dx}(f(x)) = kf'(x)$$

Thus, the derivative of a constant times a differentiable function is the product of the constant times the derivative of the function. This rule allows you to factor out constants when you are finding a derivative. The rule applies even when the constant is in the denominator as shown here:

$$\frac{d}{dx}\left(\frac{f(x)}{k}\right) = \frac{d}{dx}\left(\frac{1}{k}f(x)\right) = \frac{1}{k}\frac{d}{dx}(f(x)) = \frac{1}{k}f'(x)$$

- If $f(x) = -5x^2$, then $f'(x) = -5\frac{d}{dx}(x^2) = -5(2)x^1 = -10x$
- If $y = 6\left(x^{\frac{1}{2}}\right)$, then $y' = \frac{dy}{dx} = \frac{d}{dx}6\left(x^{\frac{1}{2}}\right) = 6\frac{d}{dx}\left(x^{\frac{1}{2}}\right) = 6\left(\frac{1}{2}\right)x^{-\frac{1}{2}} = 3x^{-\frac{1}{2}}$
- $\frac{d}{dx}(4x^{-1}) = 4\frac{d}{dx}(x^{-1}) = -4x^{-2}$

EXERCISE

**5·1**

*For problems 1–10, use the constant multiple of a function rule to find the derivative of the given function.*

1. $f(x) = 2x^3$
2. $g(x) = \frac{x^{100}}{25}$
3. $f(x) = 20x^{\frac{1}{4}}$
4. $y = -16\sqrt{x}$
5. $f(t) = \frac{2t}{3}$
6. $f(x) = \frac{x^{\pi}}{2\pi}$
7. $f(x) = \frac{10}{x^5}$
8. $s(t) = 100t^{0.6}$
9. $h(s) = -25s^{\frac{4}{5}}$
10. $f(x) = \frac{1}{4\sqrt[3]{x^2}}$

*For problems 11–15, find the indicated numerical derivative.*

11. $f'(3)$ when $f(x) = 2x^3$
12. $g'(1)$ when $g(x)=\frac{x^{100}}{25}$
13. $f'(81)$ when $f(x)=20x^{\frac{1}{4}}$
14. $\left.\frac{dy}{dx}\right|_{25}$ when $y=-16\sqrt{x}$
15. $f'(200)$ when $f(t)=\frac{2t}{3}$

# Rule for sums and differences

For all $x$ where both $f$ and $g$ are differentiable functions, the function $(f+g)$ is differentiable with its derivative given by

$$\frac{d}{dx}(f(x)+g(x))=f'(x)+g'(x)$$

Similarly, for all $x$ where both $f$ and $g$ are differentiable functions, the function $(f-g)$ is differentiable with its derivative given by

$$\frac{d}{dx}(f(x)-g(x))=f'(x)-g'(x)$$

Thus, the derivative of the sum (or difference) of two differentiable functions is equal to the sum (or difference) of the derivatives of the individual functions.

- If $h(x)=-5x^2+x$, then $h'(x)=\frac{d}{dx}(-5x^2)+\frac{d}{dx}(x)=-10x+1$
- If $y=3x^4-2x^3+5x+1$, then $y'=\frac{d}{dx}3x^4-\frac{d}{dx}2x^3+\frac{d}{dx}5x+\frac{d}{dx}1$
  $=12x^3-6x^2+5+0=12x^3-6x^2+5$
- $\frac{d}{dx}(10x^5-3x)=\frac{d}{dx}(10x^5)-\frac{d}{dx}(3x)=50x^4-3$

EXERCISE

**5·2**

*For problems 1–10, use the rule for sums and differences to find the derivative of the given function.*

1. $f(x)=x^7+2x^{10}$
2. $h(x)=30-5x^2$
3. $g(x)=x^{100}-40x^5$
4. $C(x)=1000+200x-40x^2$
5. $y=\frac{-15}{x}+25$
6. $s(t)=16t^2-\frac{2t}{3}+10$

7. $g(x)=\frac{x^{100}}{25}-20\sqrt{x}$

8. $y=12x^{0.2}+0.45x$

9. $q(v)=v^{\frac{2}{5}}+7-15v^{\frac{3}{5}}$

10. $f(x)=\frac{5}{2x^2}+\frac{5}{2x^{-2}}-\frac{5}{2}$

*For problems 11–15, find the indicated numerical derivative.*

11. $h'\left(\frac{1}{2}\right)$ when $h(x)=30-5x^2$

12. $C'(300)$ when $C(x)=1000+200x-40x^2$

13. $s'(0)$ when $s(t)=16t^2-\frac{2t}{3}+10$

14. $q'(32)$ when $q(v)=v^{\frac{2}{5}}+7-15v^{\frac{3}{5}}$

15. $f'(6)$ when $f(x)=\frac{5}{2x^2}+\frac{5}{2x^{-2}}-\frac{5}{2}$

# Product rule

For all $x$ where both $f$ and $g$ are differentiable functions, the function $(fg)$ is differentiable with its derivative given by

$$\frac{d}{dx}(f(x)g(x))=f(x)g'(x)+g(x)f'(x)$$

Thus, the derivative of the product of two differentiable functions is equal to the first function times the derivative of the second function plus the second function times the derivative of the first function.

- If $h(x)=(x^2+4)(2x-3)$,

  then $h'(x)=(x^2+4)\frac{d}{dx}(2x-3)+(2x-3)\frac{d}{dx}(x^2+4)=(x^2+4)(2)+(2x-3)(2x)$

  $=2x^2+8+4x^2-6x=6x^2-6x+8$

- If $y=(2x^3+1)(-x^2+5x+10)$,

  then $y'=(2x^3+1)\frac{d}{dx}(-x^2+5x+10)+(-x^2+5x+10)\frac{d}{dx}(2x^3+1)$

  $=(2x^3+1)(-2x+5)+(-x^2+5x+10)(6x^2)$

  $=(-4x^4+10x^3-2x+5)+(-6x^4+30x^3+60x^2)$

  $=-10x^4+40x^3+60x^2-2x+5$

Notice in the following example that converting to negative and fractional exponents makes differentiating easier.

- $\frac{d}{dx}\left[(x^2-5)\left(\frac{3}{x}+2\sqrt{x}\right)\right]=(x^2-5)\frac{d}{dx}\left(3x^{-1}+2x^{\frac{1}{2}}\right)+\left(3x^{-1}+2x^{\frac{1}{2}}\right)\frac{d}{dx}(x^2-5)$

$$=(x^2-5)\left(-3x^{-2}+x^{-\frac{1}{2}}\right)+\left(3x^{-1}+2x^{\frac{1}{2}}\right)(2x)$$

$$=-3x^0+x^{\frac{3}{2}}+15x^{-2}-5x^{-\frac{1}{2}}+6x^0+4x^{\frac{3}{2}}$$

$$=5x^{\frac{3}{2}}+15x^{-2}-5x^{-\frac{1}{2}}+3.$$

You might choose to write answers without negative or fractional exponents.

EXERCISE

**5·3**

*For problems 1–10, use the product rule to find the derivative of the given function.*

1. $f(x)=(2x^2+3)(2x-3)$
2. $h(x)=(4x^3+1)(-x^2+2x+5)$
3. $g(x)=(x^2-5)\left(\frac{3}{x}\right)$
4. $C(x)=(50+20x)(100-2x)$
5. $y=\left(\frac{-15}{\sqrt{x}}+25\right)(\sqrt{x}+5)$
6. $s(t)=\left(4t-\frac{1}{2}\right)\left(5t+\frac{3}{4}\right)$
7. $g(x)=(2x^3+2x^2)(2\sqrt[3]{x})$
8. $f(x)=\frac{10}{x^5}\cdot\frac{x^3+1}{5}$
9. $q(v)=(v^2+7)(-5v^{-2}+2)$
10. $f(x)=(2x^3+3)(3-\sqrt[3]{x^2})$

*For problems 11–15, find the indicated numerical derivative.*

11. $f'(1.5)$ when $f(x)=(2x^2+3)(2x-3)$
12. $g'(10)$ when $g(x)=(x^2-5)\left(\frac{3}{x}\right)$
13. $C'(150)$ when $C(x)=(50+20x)(100-2x)$
14. $\left.\frac{dy}{dx}\right|_{x=25}$ when $y=\left(\frac{-15}{\sqrt{x}}+25\right)(\sqrt{x}+5)$
15. $f'(2)$ when $f(x)=\frac{10}{x^5}\cdot\frac{x^3+1}{5}$

# Quotient rule

For all $x$ where both $f$ and $g$ are differentiable functions and $g(x)\neq 0$, the function $\left(\frac{f}{g}\right)$ is differentiable with its derivative given by

$$\frac{d}{dx}\left(\frac{f(x)}{g(x)}\right)=\frac{g(x)f'(x)-f(x)g'(x)}{(g(x))^2},\, g(x)\neq 0$$

Thus, the derivative of the quotient of two differentiable functions is equal to the denominator function times the derivative of the numerator function minus the numerator function times the derivative of the denominator function all divided by the square of the denominator function, for all real numbers $x$ for which the denominator function is not equal to zero.

- If $h(x)=\frac{-5x^2+4}{3x}$, then $h'(x)=\frac{(3x)\frac{d}{dx}(-5x^2+4)-(-5x^2+4)\frac{d}{dx}(3x)}{(3x)^2}$

$$=\frac{(3x)(-10x)-(-5x^2+4)(3)}{(3x)^2}=\frac{-30x^2+15x^2-12}{9x^2}$$

$$=\frac{-15x^2-12}{9x^2}=-\frac{5x^2+4}{3x^2}$$

- If $y=\frac{1}{\sqrt{x}}$, then $y'=\frac{(\sqrt{x})\frac{d}{dx}(1)-(1)\frac{d}{dx}(\sqrt{x})}{(\sqrt{x})^2}=\frac{(\sqrt{x})(0)-(1)\frac{d}{dx}\left(x^{\frac{1}{2}}\right)}{(\sqrt{x})^2}$

$$=\frac{-(1)\frac{1}{2}\left(x^{-\frac{1}{2}}\right)}{x}=-\frac{1}{2x^{\frac{3}{2}}}$$

- $\frac{d}{dx}\left(\frac{8x^{\frac{5}{4}}}{2x^4+6}\right)=\frac{(2x^4+6)\frac{d}{dx}\left(8x^{\frac{5}{4}}\right)-\left(8x^{\frac{5}{4}}\right)\frac{d}{dx}(2x^4+6)}{(2x^4+6)^2}=\frac{(2x^4+6)\left(10x^{\frac{1}{4}}\right)-\left(8x^{\frac{5}{4}}\right)(8x^3)}{(2x^4+6)^2}$

$$=\frac{\left(20x^{\frac{17}{4}}+60x^{\frac{1}{4}}\right)-\left(64x^{\frac{17}{4}}\right)}{4x^8+24x^4+36}=\frac{20x^{\frac{17}{4}}+60x^{\frac{1}{4}}-64x^{\frac{17}{4}}}{4x^8+24x^4+36}=\frac{15x^{\frac{1}{4}}-11x^{\frac{17}{4}}}{x^8+6x^4+9}$$

## EXERCISE 5·4

*For problems 1–10, use the quotient rule to find the derivative of the given function.*

1. $f(x)=\frac{5x+2}{3x-1}$
2. $h(x)=\frac{4-5x^2}{8x}$
3. $g(x)=\frac{5}{\sqrt{x}}$
4. $f(x)=\frac{3x^{\frac{3}{2}}-1}{2x^{\frac{1}{2}}+6}$
5. $y=\frac{-15}{x}$
6. $s(t)=\frac{2t^{\frac{3}{2}}-3}{4t^{\frac{1}{2}}+6}$
7. $g(x)=\frac{x^{100}}{x^{-5}+10}$
8. $y=\frac{4-5x^3}{8x^2-7}$
9. $q(v)=\frac{v^3+2}{v^2-\frac{1}{v^3}}$
10. $f(x)=\frac{-4x^2}{\frac{4}{x^2}+8}$

*For problems 11–15, find the indicated numerical derivative.*

11. $f'(25)$ when $f(x) = \dfrac{5x+2}{3x-1}$

12. $h'(0.2)$ when $h(x) = \dfrac{4-5x^2}{8x}$

13. $g'(0.25)$ when $g(x) = \dfrac{5}{\sqrt{x}}$

14. $\left.\dfrac{dy}{dx}\right|_{10}$ when $y = \dfrac{-15}{x}$

15. $g'(1)$ when $g(x) = \dfrac{x^{100}}{x^{-5}+10}$

# Chain rule

If $y = f(u)$ and $u = g(x)$ are differentiable functions of $u$ and $x$, respectively, then the composition of $f$ and $g$, defined by $y = f(g(x))$, is differentiable with its derivative given by

$$\frac{dy}{dx} = \frac{dy}{du} \cdot \frac{du}{dx}$$

or equivalently,

$$\frac{d}{dx}[f(g(x))] = f'(g(x))g'(x)$$

Notice that $y = f(g(x))$ is a "function of a function of $x$"; that is, $f$'s argument is the function denoted by $g(x)$, which itself is a function of $x$. Thus, to find $\frac{d}{dx}[f(g(x))]$, you must differentiate $f$ with respect to $g(x)$ first, and then multiply the result by the derivative of $g(x)$ with respect to $x$.

The examples that follow illustrate the chain rule.

- Find $y'$, when $y = \sqrt{3x^4 - 2x^3 + 5x + 1}$; let $u = 3x^4 - 2x^3 + 5x + 1$,

  then $y' = \dfrac{dy}{dx} = \dfrac{dy}{du} \cdot \dfrac{du}{dx} = \dfrac{d}{du}(u)^{\frac{1}{2}} \cdot \dfrac{d}{dx}(3x^4 - 2x^3 + 5x + 1) = \dfrac{1}{2}u^{-\frac{1}{2}} \cdot (12x^3 - 6x^2 + 5)$

  $= \dfrac{1}{2}(3x^4 - 2x^3 + 5x + 1)^{-\frac{1}{2}} \cdot (12x^3 - 6x^2 + 5) = \dfrac{12x^3 - 6x^2 + 5}{2\sqrt{3x^4 - 2x^3 + 5x + 1}}$

- Find $f'(x)$, when $f(x) = (x^2 - 8)^3$; let $g(x) = x^2 - 8$,

  then $\dfrac{d}{dx}[f(g(x))] = \dfrac{d}{dx}[(x^2 - 8)^3] = f'(g(x))g'(x)$

  $= 3(g(x))^2 g'(x) = 3(x^2 - 8)^2 \cdot 2x = 6x(x^2 - 8)^2$

- $\dfrac{d}{dx}(\sqrt{x}+1)^4 = 4(\sqrt{x}+1)^3 \dfrac{d}{dx}(\sqrt{x}+1) = 4(\sqrt{x}+1)^3\left(\dfrac{1}{2}x^{-\frac{1}{2}}\right) = \dfrac{2(\sqrt{x}+1)^3}{\sqrt{x}}$

EXERCISE 5·5

*For problems 1–10, use the chain rule to find the derivative of the given function.*

1. $f(x)=(3x^2-10)^3$
2. $g(x)=40(3x^2-10)^3$
3. $h(x)=10(3x^2-10)^{-3}$
4. $h(x)=(\sqrt{x}+3)^2$
5. $f(u)=\left(\dfrac{1}{u^2}-u\right)^3$
6. $y=\dfrac{1}{(x^2-8)^3}$
7. $y=\sqrt{2x^3+5x+1}$
8. $s(t)=(2t^3+5t)^{\frac{1}{3}}$
9. $f(x)=\dfrac{10}{(2x-6)^5}$
10. $C(t)=\dfrac{50}{\sqrt{15t+120}}$

*For problems 11–15, find the indicated numerical derivative.*

11. $f'(10)$ when $f(x)=(3x^2-10)^3$
12. $h'(3)$ when $h(x)=10(3x^2-10)^{-3}$
13. $f'(144)$ when $f(x)=(\sqrt{x}+3)^2$
14. $f'(2)$ when $f(u)=\left(\dfrac{1}{u^2}-u\right)^3$
15. $\left.\dfrac{dy}{dx}\right|_4$ when $y=\dfrac{1}{(x^2-8)^3}$

# Implicit differentiation

Thus far, you've seen how to find the derivative of a function only if the function is expressed in what is called explicit form. A function in **explicit form** is defined by an equation of the type $y=f(x)$, where $y$ is on one side of the equation and all the terms containing $x$ are on the other side. For example, the function $f$ defined by $y=f(x)=x^3+5$ is expressed in explicit form. For this function the variable $y$ is defined **explicitly** as a function of the variable $x$.

On the other hand, for equations in which the variables $x$ and $y$ appear on the same side of the equation, the function is said to be expressed in **implicit form**. For example, the equation $x^2y=1$ defines the function $y=\dfrac{1}{x^2}$ **implicitly** in terms of $x$. In this case, the implicit form of the equation can be solved for $y$ as a function of $x$; however, for many implicit forms, it is difficult and sometimes impossible to solve for $y$ in terms of $x$.

Under the assumption that $\dfrac{dy}{dx}$, the derivative of $y$ with respect to $x$, exists, you can use the technique of **implicit differentiation** to find $\dfrac{dy}{dx}$ when a function is expressed in implicit form—regardless of whether you can express the function in explicit form. Use the following steps:

1. Differentiate every term on both sides of the equation *with respect to x*.
2. Solve the resulting equation for $\dfrac{dy}{dx}$.

**PROBLEM** Given the equation $x^2+2y^3=30$, use implicit differentiation to find $\dfrac{dy}{dx}$.

**SOLUTION** Step 1: Differentiate every term on both sides of the equation *with respect to x*:

$$\frac{d}{dx}(x^2+2y^3)=\frac{d}{dx}(30)$$

$$\frac{d}{dx}(x^2)+\frac{d}{dx}(2y^3)=\frac{d}{dx}(30)$$

$$2x+6y^2\frac{dy}{dx}=0$$

Step 2: Solve the resulting equation for $\dfrac{dy}{dx}$.

$$6y^2\frac{dy}{dx}=-2x$$

$$\frac{dy}{dx}=\frac{-2x}{6y^2}$$

Note that in this example, $\dfrac{dy}{dx}$ is expressed in terms of both $x$ and $y$. To evaluate such a derivative, you would need to know both $x$ and $y$ at a particular point $(x, y)$. You can denote the numerical derivative as $\left.\dfrac{dy}{dx}\right|_{(x,y)}$.

The example that follows illustrates this situation.

$\dfrac{dy}{dx}=\dfrac{-2x}{6y^2}$ at (3, 1) is given by

$$\left.\frac{dy}{dx}\right|_{(3,1)}=\left.\frac{-2x}{6y^2}\right|_{(3,1)}=\frac{-2(3)}{6(1)^2}=-1$$

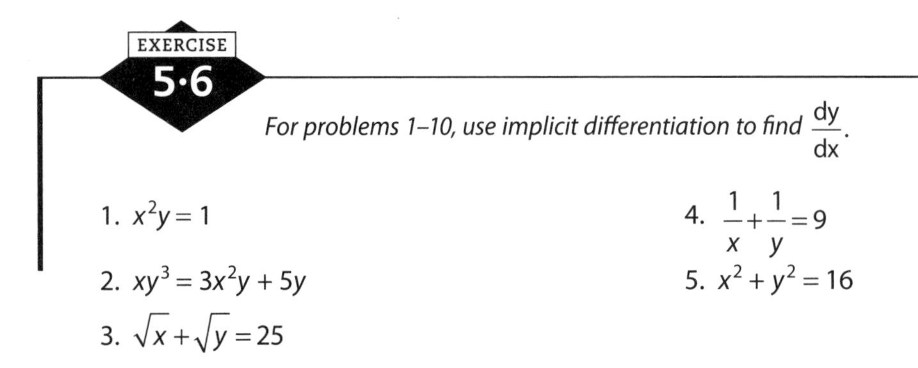

EXERCISE 5·6

*For problems 1–10, use implicit differentiation to find* $\dfrac{dy}{dx}$.

1. $x^2y=1$
2. $xy^3=3x^2y+5y$
3. $\sqrt{x}+\sqrt{y}=25$
4. $\dfrac{1}{x}+\dfrac{1}{y}=9$
5. $x^2+y^2=16$

*For problems 6–10, find the indicated numerical derivative.*

6. $\left.\frac{dy}{dx}\right|_{(3,1)}$ when $x^2y = 1$

7. $\left.\frac{dy}{dx}\right|_{(5,2)}$ when $xy^3 = 3x^2y + 5y$

8. $\left.\frac{dy}{dx}\right|_{(4,9)}$ when $\sqrt{x} + \sqrt{y} = 25$

9. $\left.\frac{dy}{dx}\right|_{(5,10)}$ when $\frac{1}{x} + \frac{1}{y} = 9$

10. $\left.\frac{dy}{dx}\right|_{(2,1)}$ when $x^2 + y^2 = 16$

# Additional derivatives

## Derivative of the natural exponential function $e^x$

Exponential functions are defined by equations of the form $y = f(x) = b^x$ $(b \neq 1, b > 0)$, where $b$ is the base of the exponential function. The natural exponential function is the exponential function whose base is the irrational number $e$. The number $e$ is the limit as $n$ approaches infinity of $\left(1+\frac{1}{n}\right)^n$, which is approximately 2.718281828 (to nine decimal places).

The natural exponential function is its own derivative; that is, $\frac{d}{dx}(e^x) = e^x$.

Furthermore, by the chain rule, if $u$ is a differentiable function of $x$, then

$$\frac{d}{dx}(e^u) = e^u \cdot \frac{du}{dx}$$

- If $f(x) = 6e^x$, then $f'(x) = 6 \cdot \frac{d}{dx}(e^x) = 6e^x$
- If $y = e^{2x}$, then $y' = e^{2x} \cdot \frac{d}{dx}(2x) = e^{2x}(2) = 2e^{2x}$
- $\frac{d}{dx}(e^{-3x^2}) = e^{-3x^2} \cdot \frac{d}{dx}(-3x^2) = e^{-3x^2}(-6x) = -6xe^{-3x^2}$

EXERCISE 6·1

*Find the derivative of the given function.*

1. $f(x) = 20e^x$
2. $y = e^{3x}$
3. $g(x) = e^{5x^3}$
4. $y = -4e^{5x^3}$
5. $h(x) = e^{-10x^3}$
6. $f(x) = 15x^2 + 10e^x$
7. $g(x) = e^{7x-2x^3}$
8. $f(t) = \frac{100}{e^{-0.5t}}$
9. $g(t) = 2500e^{2t+1}$
10. $f(x) = \frac{1}{\sqrt{2\pi}}e^{-\frac{x^2}{2}}$

# Derivative of the natural logarithmic function $\ln x$

Logarithmic functions are defined by equations of the form $y = f(x) = \log_b x$ if and only if $b^y = x\,(x > 0)$, where $b$ is the base of the logarithmic function, $(b \neq 1, b > 0)$. For a given base, the logarithmic function is the inverse function of the corresponding exponential function, and reciprocally. The logarithmic function defined by $y = \log_e x$, usually denoted $\ln x$, is the natural logarithmic function. It is the inverse function of the natural exponential function $y = e^x$.

The derivative of the natural logarithmic function is as follows:

$$\frac{d}{dx}(\ln x) = \frac{1}{x}$$

Furthermore, by the chain rule, if $u$ is a differentiable function of $x$, then

$$\frac{d}{dx}(\ln u) = \frac{1}{u} \cdot \frac{du}{dx}$$

- If $f(x) = 6\ln x$, then $f'(x) = 6 \cdot \frac{d}{dx}(\ln x) = 6 \cdot \frac{1}{x} = \frac{6}{x}$
- If $y = \ln(2x^3)$, then $y' = \frac{1}{2x^3} \cdot \frac{d}{dx}(2x^3) = \frac{1}{2x^3} \cdot (6x^2) = \frac{3}{x}$
- $\frac{d}{dx}(\ln 2x) = \frac{1}{2x} \cdot \frac{d}{dx}(2x) = \frac{1}{2x} \cdot (2) = \frac{1}{x}$

The above example illustrates that for any nonzero constant $k$,

$$\frac{d}{dx}(\ln kx) = \frac{1}{kx} \cdot \frac{d}{dx}(kx) = \frac{1}{kx} \cdot (k) = \frac{1}{x}$$

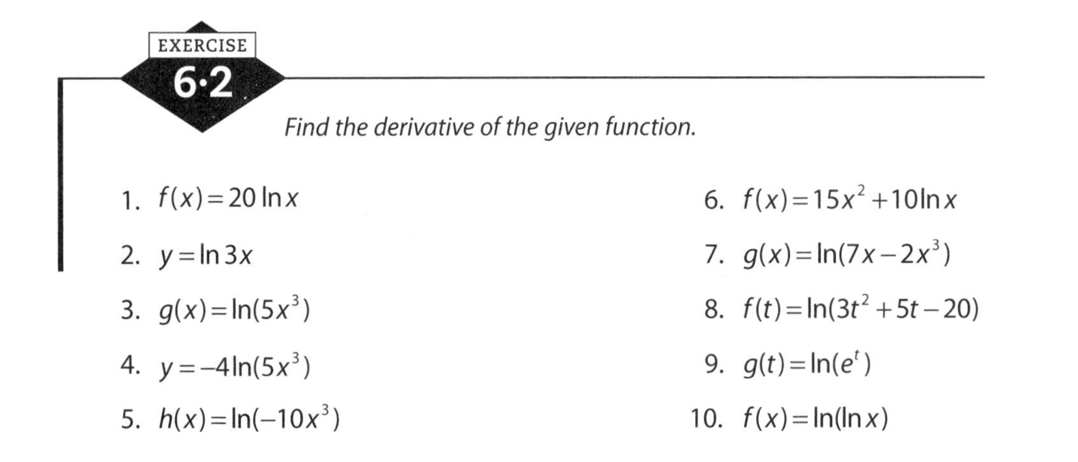

EXERCISE 6·2

*Find the derivative of the given function.*

1. $f(x) = 20\ln x$
2. $y = \ln 3x$
3. $g(x) = \ln(5x^3)$
4. $y = -4\ln(5x^3)$
5. $h(x) = \ln(-10x^3)$
6. $f(x) = 15x^2 + 10\ln x$
7. $g(x) = \ln(7x - 2x^3)$
8. $f(t) = \ln(3t^2 + 5t - 20)$
9. $g(t) = \ln(e^t)$
10. $f(x) = \ln(\ln x)$

# Derivatives of exponential functions for bases other than $e$

Suppose $b$ is a positive real number $(b \neq 1)$, then

$$\frac{d}{dx}(b^x) = (\ln b)b^x$$

Furthermore, by the chain rule, if $u$ is a differentiable function of $x$, then

$$\frac{d}{dx}(b^u) = (\ln b)b^u \cdot \frac{du}{dx}$$

- If $f(x) = (6)2^x$, then $f'(x) = 6 \cdot \frac{d}{dx}(2^x) = 6(\ln 2)2^x$
- If $y = 5^{2x}$, then $y' = (\ln 5)5^{2x} \cdot \frac{d}{dx}(2x) = (\ln 5)5^{2x} \cdot (2) = 2(\ln 5)5^{2x}$
- $\frac{d}{dx}(10^{-3x^2}) = (\ln 10)10^{-3x^2} \cdot \frac{d}{dx}(-3x^2) = (\ln 10)10^{-3x^2}(-6x) = -6x(\ln 10)10^{-3x^2}$

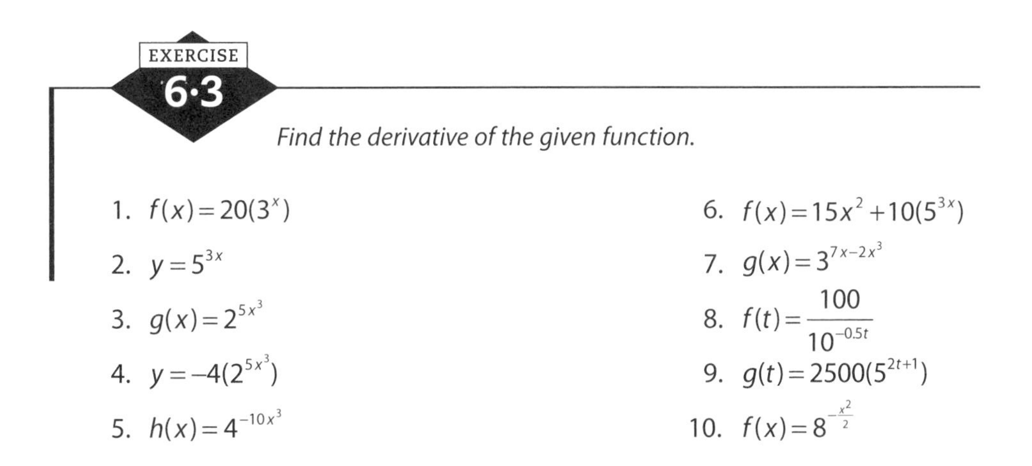

EXERCISE 6·3

*Find the derivative of the given function.*

1. $f(x) = 20(3^x)$
2. $y = 5^{3x}$
3. $g(x) = 2^{5x^3}$
4. $y = -4(2^{5x^3})$
5. $h(x) = 4^{-10x^3}$
6. $f(x) = 15x^2 + 10(5^{3x})$
7. $g(x) = 3^{7x-2x^3}$
8. $f(t) = \frac{100}{10^{-0.5t}}$
9. $g(t) = 2500(5^{2t+1})$
10. $f(x) = 8^{-\frac{x^2}{2}}$

# Derivatives of logarithmic functions for bases other than *e*

Suppose $b$ is a positive real number ($b \neq 1$), then

$$\frac{d}{dx}(\log_b x) = \frac{1}{(\ln b)x}$$

Furthermore, by the chain rule, if $u$ is a differentiable function of $x$, then

$$\frac{d}{dx}(\log_b u) = \frac{1}{(\ln b)u} \cdot \frac{du}{dx}$$

- If $f(x) = 6\log_2 x$, then $f'(x) = 6 \cdot \frac{d}{dx}(\log_2 x) = 6 \cdot \frac{1}{(\ln 2)x} = \frac{6}{x \ln 2}$
- If $y = \log_5(2x^3)$, then $y' = \frac{1}{(\ln 5)2x^3} \cdot \frac{d}{dx}(2x^3) = \frac{1}{(\ln 5)2x^3} \cdot (6x^2) = \frac{3}{x \ln 5}$
- $\frac{d}{dx}(\log_3 2x) = \frac{1}{(\ln 3)2x} \cdot \frac{d}{dx}(2x) = \frac{1}{(\ln 3)2x} \cdot (2) = \frac{1}{x \ln 3}$

The above example illustrates that for any nonzero constant $k$,

$$\frac{d}{dx}(\log_b kx) = \frac{1}{(\ln b)kx} \cdot \frac{d}{dx}(kx) = \frac{1}{(\ln b)kx} \cdot (k) = \frac{1}{x \ln b}$$

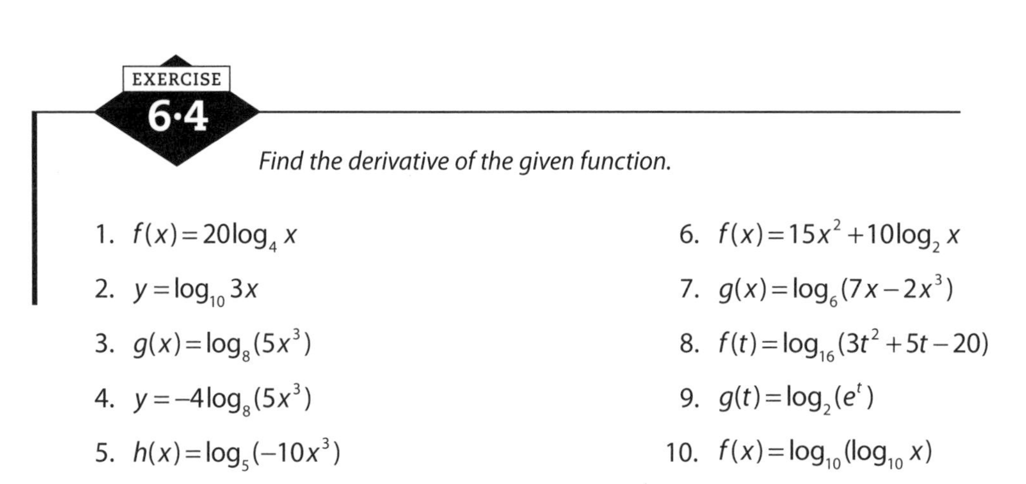

EXERCISE 6·4

*Find the derivative of the given function.*

1. $f(x) = 20\log_4 x$
2. $y = \log_{10} 3x$
3. $g(x) = \log_8(5x^3)$
4. $y = -4\log_8(5x^3)$
5. $h(x) = \log_5(-10x^3)$
6. $f(x) = 15x^2 + 10\log_2 x$
7. $g(x) = \log_6(7x - 2x^3)$
8. $f(t) = \log_{16}(3t^2 + 5t - 20)$
9. $g(t) = \log_2(e^t)$
10. $f(x) = \log_{10}(\log_{10} x)$

# Derivatives of trigonometric functions

The derivatives of the trigonometric functions are as follows:

- $\frac{d}{dx}(\sin x) = \cos x$
- $\frac{d}{dx}(\cos x) = -\sin x$
- $\frac{d}{dx}(\tan x) = \sec^2 x$
- $\frac{d}{dx}(\cot x) = -\csc^2 x$
- $\frac{d}{dx}(\sec x) = \sec x \tan x$
- $\frac{d}{dx}(\csc x) = -\csc x \cot x$

Furthermore, by the chain rule, if $u$ is a differentiable function of $x$, then

- $\frac{d}{dx}(\sin u) = \cos u \cdot \frac{du}{dx}$
- $\frac{d}{dx}(\cos u) = -\sin u \cdot \frac{du}{dx}$
- $\frac{d}{dx}(\tan u) = \sec^2 u \cdot \frac{du}{dx}$

- $\frac{d}{dx}(\cot u) = -\csc^2 u \cdot \frac{du}{dx}$
- $\frac{d}{dx}(\sec u) = (\sec u \tan u) \cdot \frac{du}{dx}$
- $\frac{d}{dx}(\csc u) = (-\csc u \cot u) \cdot \frac{du}{dx}$
- If $h(x) = \sin 3x$, then $h'(x) = (\cos 3x)\frac{d}{dx}(3x) = (\cos 3x)(3) = 3\cos 3x$
- If $y = 3\cos\left(\frac{x}{3}\right)$, then $y' = -3\sin\left(\frac{x}{3}\right)\frac{d}{dx}\left(\frac{x}{3}\right) = -3\left[\sin\left(\frac{x}{3}\right)\right]\left(\frac{1}{3}\right) = -\sin\left(\frac{x}{3}\right)$
- $\frac{d}{dx}(\tan 2x + \cot 2x) = \frac{d}{dx}(\tan 2x) + \frac{d}{dx}(\cot 2x) = \sec^2(2x)\frac{d}{dx}(2x) - \csc^2(2x)\frac{d}{dx}(2x)$

$$= [\sec^2(2x)](2) - [\csc^2(2x)](2) = 2\sec^2(2x) - 2\csc^2(2x)$$

EXERCISE 6·5

*Find the derivative of the given function.*

1. $f(x) = 5\sin 3x$
2. $h(x) = \frac{1}{4}\cos(2x^2)$
3. $g(x) = 5\tan\left(\frac{3x}{5}\right)$
4. $f(x) = 10\sec 2x$
5. $y = \frac{2}{3}\sec(2x^3)$
6. $s(t) = 4\cot 5t$
7. $g(x) = 6\tan^3\left(\frac{2x}{3}\right) - 20\sqrt{x}$
8. $f(x) = 2x\sin x + \cos 2x$
9. $h(x) = \frac{\sin 3x}{1 + \sin 3x}$
10. $f(x) = e^{4x}\sin 2x$

# Derivatives of inverse trigonometric functions

The derivatives of the inverse trigonometric functions are as follows:

- $\frac{d}{dx}(\sin^{-1} x) = \frac{1}{\sqrt{1 - x^2}}$
- $\frac{d}{dx}(\cos^{-1} x) = \frac{-1}{\sqrt{1 - x^2}}$
- $\frac{d}{dx}(\tan^{-1} x) = \frac{1}{1 + x^2}$

- $\dfrac{d}{dx}(\cot^{-1} x) = \dfrac{-1}{1+x^2}$
- $\dfrac{d}{dx}(\sec^{-1} x) = \dfrac{1}{|x|\sqrt{x^2-1}}$
- $\dfrac{d}{dx}(\csc^{-1} x) = \dfrac{-1}{|x|\sqrt{x^2-1}}$

Furthermore, by the chain rule, if $u$ is a differentiable function of $x$, then

- $\dfrac{d}{dx}(\sin^{-1} u) = \dfrac{1}{\sqrt{1-u^2}} \cdot \dfrac{du}{dx}$
- $\dfrac{d}{dx}(\cos^{-1} u) = \dfrac{-1}{\sqrt{1-u^2}} \cdot \dfrac{du}{dx}$
- $\dfrac{d}{dx}(\tan^{-1} u) = \dfrac{1}{1+u^2} \cdot \dfrac{du}{dx}$
- $\dfrac{d}{dx}(\cot^{-1} u) = \dfrac{-1}{1+u^2} \cdot \dfrac{du}{dx}$
- $\dfrac{d}{dx}(\sec^{-1} u) = \dfrac{1}{|u|\sqrt{u^2-1}} \cdot \dfrac{du}{dx}$
- $\dfrac{d}{dx}(\csc^{-1} u) = \dfrac{-1}{|u|\sqrt{u^2-1}} \cdot \dfrac{du}{dx}$
- If $h(x) = \sin^{-1}(2x)$, then $h'(x) = \dfrac{1}{\sqrt{1-(2x)^2}} \cdot \dfrac{d}{dx}(2x) = \dfrac{1}{\sqrt{1-4x^2}} \cdot (2) = \dfrac{2}{\sqrt{1-4x^2}}$
- If $y = \cos^{-1}\left(\dfrac{x}{3}\right)$, then $y' = \dfrac{-1}{\sqrt{1-\left(\dfrac{x}{3}\right)^2}} \cdot \dfrac{d}{dx}\left(\dfrac{x}{3}\right) = \dfrac{-1}{\sqrt{1-\dfrac{x^2}{9}}} \cdot \left(\dfrac{1}{3}\right) = -\dfrac{1}{3\sqrt{\dfrac{9-x^2}{9}}}$

$$= -\frac{1}{3\left(\frac{1}{3}\right)\sqrt{9-x^2}} = -\frac{1}{\sqrt{9-x^2}}$$

- $\dfrac{d}{dx}(\tan^{-1} x + \cot^{-1} x) = \dfrac{d}{dx}(\tan^{-1} x) + \dfrac{d}{dx}(\cot^{-1} x) = \dfrac{1}{1+x^2} + \dfrac{-1}{1+x^2} = 0$

**Note:** An alternative notation for an inverse trigonometric function is to prefix the original function with "arc," as in "arcsin $x$," which is read "arcsine of $x$" or "an angle whose sine is $x$." An advantage of this notation is that it helps you avoid the common error of confusing the inverse function; for example, $\sin^{-1} x$, with its reciprocal $(\sin x)^{-1} = \dfrac{1}{\sin x}$.

EXERCISE 6·6

*Find the derivative of the given function.*

1. $f(x) = \sin^{-1}(-x^3)$
2. $h(x) = \cos^{-1}(e^x)$
3. $g(x) = \tan^{-1}(x^2)$
4. $f(x) = \cot^{-1}(7x - 5)$
5. $y = \frac{1}{15}\sin^{-1}(5x^3)$
6. $f(x) = \cos^{-1}(x^2)$
7. $h(x) = \csc^{-1}(2x)$
8. $g(x) = 4\sec^{-1}\left(\frac{x}{2}\right)$
9. $f(x) = x\sin^{-1}(7x^2)$
10. $y = \arcsin(\sqrt{1 - x^2})$

# Higher-order derivatives

For a given function $f$, higher-order derivatives of $f$, if they exist, are obtained by differentiating $f$ successively multiple times. The derivative $f'$ is called the **first derivative** of $f$. The derivative of $f'$ is called the **second derivative** of $f$ and is denoted $f''$. Similarly, the derivative of $f''$ is called the **third derivative** of $f$ and is denoted $f'''$, and so on.

Other common notations for higher-order derivatives are the following:

- 1st derivative: $f'(x), y', \frac{dy}{dx}, D_x[f(x)]$
- 2nd derivative: $f''(x), y'', \frac{d^2y}{d^2x}, D_x^2[f(x)]$
- 3rd derivative: $f'''(x), y''', \frac{d^3y}{d^3x}, D_x^3[f(x)]$
- 4th derivative: $f^{(4)}(x), y^{(4)}, \frac{d^4y}{d^4x}, D_x^4[f(x)]$
- $n$th derivative: $f^{(n)}(x), y^{(n)}, \frac{d^ny}{d^nx}, D_x^n[f(x)]$

**Note:** The $n^{\text{th}}$ derivative is also called the $n^{\text{th}}$-order derivative. Thus, the first derivative is the first-order derivative; the second derivative, the second-order derivative; the third derivative, the third-order derivative; and so on.

PROBLEM Find the first three derivatives of $f$ if $f(x) = x^{100} - 40x^5$.

SOLUTION

$$f'(x) = 100x^{99} - 200x^4$$

$$f''(x) = 9900x^{98} - 800x^3$$

$$f'''(x) = 970200x^{97} - 2400x^2$$

EXERCISE

6·7

*Find the indicated derivative of the given function.*

1. If $f(x) = x^7 + 2x^{10}$, find $f'''(x)$.
2. If $h(x) = \sqrt[3]{x}$, find $h''(x)$.
3. If $g(x) = 2x$, find $g^{(5)}(x)$.
4. If $f(x) = 5e^x$, find $f^{(4)}(x)$.
5. If $y = \sin 3x$, find $\dfrac{d^3y}{d^3x}$.
6. If $s(t) = 16t^2 - \dfrac{2t}{3} + 10$, find $s''(t)$.
7. If $g(x) = \ln 3x$, find $D_x^3[g(x)]$.
8. If $f(x) = \dfrac{10}{x^5} + \dfrac{x^3}{5}$, find $f^{(4)}(x)$.
9. If $f(x) = 3^{2x}$, find $f'''(x)$.
10. If $y = \log_2 5x$, find $\dfrac{d^4y}{d^4x}$.

# INTEGRATION

Fundamentally, integration is the process of reversing the results of differentiation. In Part III you will be working with integration formulas for certain basic function types, along with practicing various techniques of integration. The material begins with a focus on indefinite integrals, and then moves on to definite integrals and the crowning triumph of integral calculus, the First Fundamental Theorem of Calculus. The highly useful Second Fundamental Theorem of Calculus and Mean Value Theorem for Integrals are also presented.

# Indefinite integral and basic integration formulas and rules

## Antiderivatives and the indefinite integral

An **antiderivative of a function *f*** on an interval $I$ is any function $F$ such that $F'(x) = \frac{d}{dx}[F(x)] = f(x)$ for all $x$ in $I$.

Thus, an antiderivative is the result of reversing the process of differentiation, so to speak. The following examples illustrate this concept.

- $5x^3$ is an antiderivative of $15x^2$ because $\frac{d}{dx}(5x^3) = 15x^2$.
- $5x^3 - 20$ is an antiderivative of $15x^2$ because $\frac{d}{dx}(5x^3 - 20) = 15x^2 - 0 = 15x^2$.
- $5x^3 + 100$ is an antiderivative of $15x^2$ because $\frac{d}{dx}(5x^3 + 100) = 15x^2 + 0 = 15x^2$.
- $\tan x$ is an antiderivative of $\sec^2 x$ because $\frac{d}{dx}(\tan x) = \sec^2 x$.
- $\tan x + 4$ is an antiderivative of $\sec^2 x$ because $\frac{d}{dx}(\tan x + 4) = \sec^2 x + 0 = \sec^2 x$.
- $\tan x - 30$ is an antiderivative of $\sec^2 x$ because $\frac{d}{dx}(\tan x - 30) = \sec^2 x - 0 = \sec^2 x$.

From these examples, you can see that, although functions have at most one derivative, they may have many (in fact, infinitely many) antiderivatives. Thus, if $F$ is an antiderivative of a function $f$ over an interval $I$, then $F(x) + C$ represents the set of antiderivatives of $f$, where $C$ is an arbitrary constant.

The **indefinite integral of *f*** is the set of all antiderivatives of $f$ and is denoted by $\int f(x)dx$. Thus,

$$\int f(x)dx = F(x) + C,$$

where $F$ is an antiderivative of $f$ over an interval $I$ and $C$ is an arbitrary constant. The process of determining the indefinite integral is called **integration**. The expression $\int f(x)dx$ is read "**the integral of *f* of *x* with respect to *x***"; $f(x)$ is called the **integrand**, $dx$ is called the **differential**, and $C$ is called the **constant of integration**.

**Note:** The differential $dx$ indicates that the integration takes place with respect to the variable $x$.

Hereafter, it will be understood that in the expression $\int f(x)\,dx = F(x) + C$, $F$ is an antiderivative of $f$ over an interval.

You likely have surmised that integration "undoes" the process of differentiation to within a constant value. In a like manner, differentiation "undoes" the process of integration. The following examples illustrate this inverse ("undoing") relationship between integration and differentiation.

PROBLEM Verify that $\int 15x^2 dx = 5x^3 + C$ by differentiating the right member.

SOLUTION $\dfrac{d}{dx}(5x^3 + C) = 15x^2 + 0 = 15x^2$

PROBLEM Verify that $\int \sec^2 x\,dx = \tan x + C$ by differentiating the right member.

SOLUTION $\dfrac{d}{dx}(\tan x + C) = \sec^2 x + 0 = \sec^2 x$

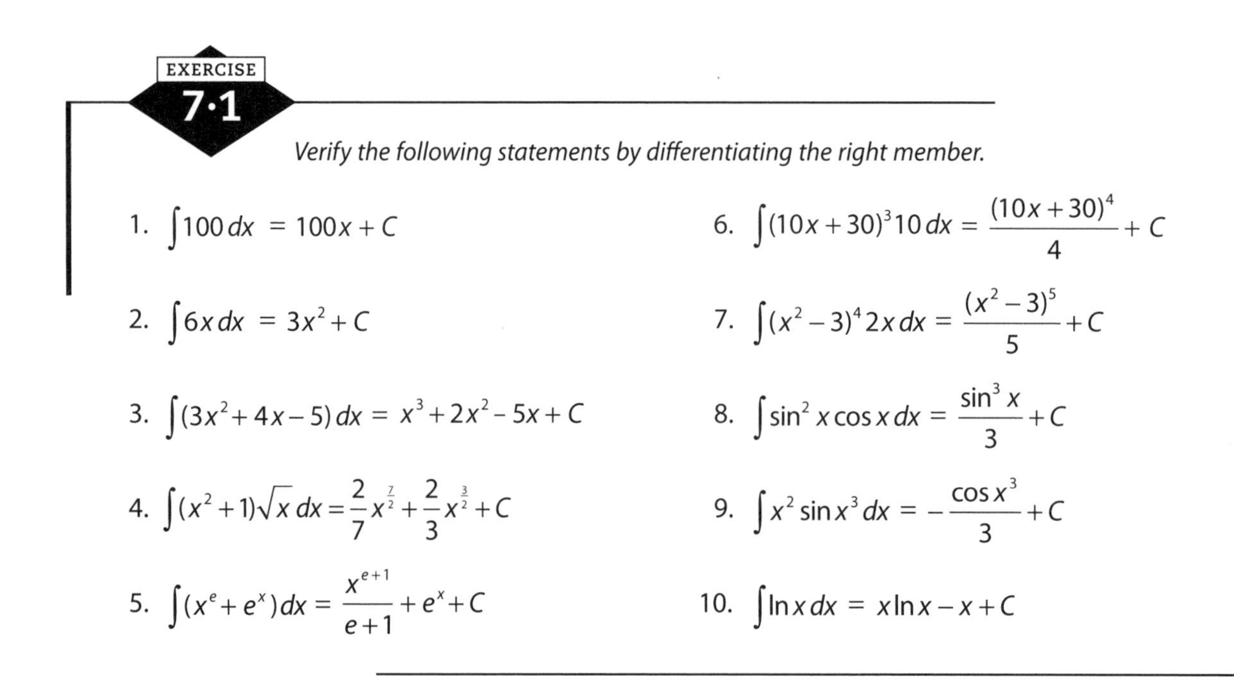

EXERCISE 7·1

*Verify the following statements by differentiating the right member.*

1. $\int 100\,dx = 100x + C$
2. $\int 6x\,dx = 3x^2 + C$
3. $\int (3x^2 + 4x - 5)\,dx = x^3 + 2x^2 - 5x + C$
4. $\int (x^2 + 1)\sqrt{x}\,dx = \frac{2}{7}x^{\frac{7}{2}} + \frac{2}{3}x^{\frac{3}{2}} + C$
5. $\int (x^e + e^x)\,dx = \dfrac{x^{e+1}}{e+1} + e^x + C$
6. $\int (10x + 30)^3 10\,dx = \dfrac{(10x + 30)^4}{4} + C$
7. $\int (x^2 - 3)^4 2x\,dx = \dfrac{(x^2 - 3)^5}{5} + C$
8. $\int \sin^2 x \cos x\,dx = \dfrac{\sin^3 x}{3} + C$
9. $\int x^2 \sin x^3\,dx = -\dfrac{\cos x^3}{3} + C$
10. $\int \ln x\,dx = x\ln x - x + C$

# Integration of constant functions

If $k$ is any constant, then $\int k\,dx = kx + C$, where $C$ is an arbitrary constant.

- $\int 3\,dx = 3x + C$
- $\int \sqrt{7}\,dx = \sqrt{7}x + C$
- $\int dx = \int 1\,dx = 1x + C = x + C$

**Note:** This solution is usually written $\int dx = x + C$.

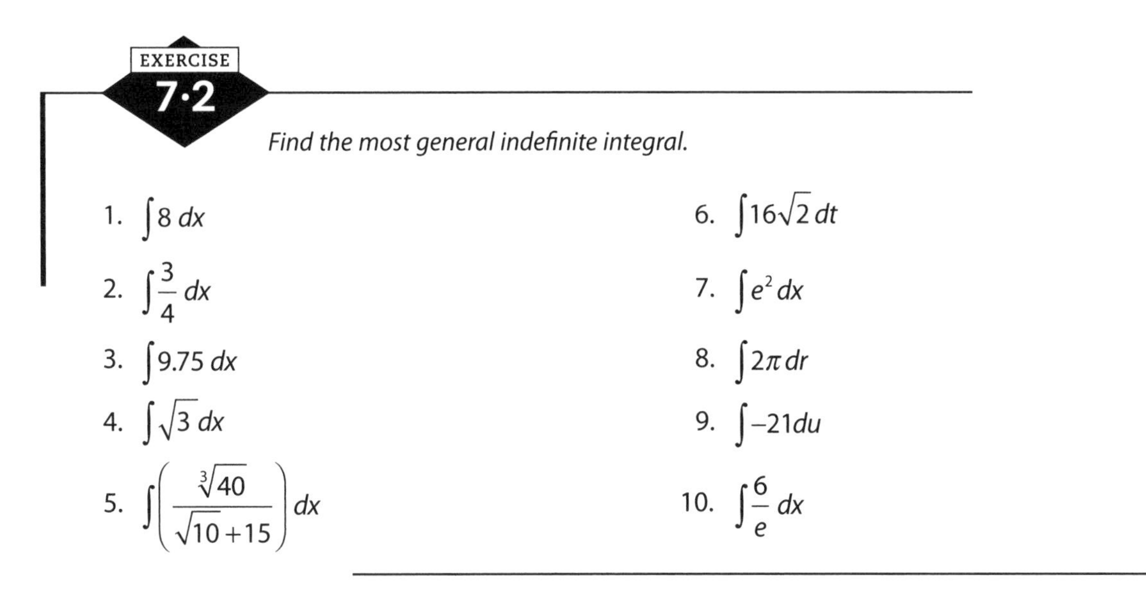
EXERCISE 7·2

*Find the most general indefinite integral.*

1. $\int 8\,dx$
2. $\int \frac{3}{4}\,dx$
3. $\int 9.75\,dx$
4. $\int \sqrt{3}\,dx$
5. $\int \left(\frac{\sqrt[3]{40}}{\sqrt{10}+15}\right)dx$
6. $\int 16\sqrt{2}\,dt$
7. $\int e^2\,dx$
8. $\int 2\pi\,dr$
9. $\int -21du$
10. $\int \frac{6}{e}\,dx$

# Integration of power functions

The following integral formulas for power functions can be derived from the formulas for differentiating power functions (see Chapter 4) and the natural logarithm function (see Chapter 6):

$$\int x^n\,dx = \frac{x^{n+1}}{n+1} + C, \text{ for all } n \neq -1;$$

and

$$\int x^{-1}dx = \int \frac{1}{x}\,dx = \ln|x| + C,$$

where $C$ is an arbitrary constant.

- $\int x^2\,dx = \frac{x^3}{3} + C$
- $\int \sqrt{x}\,dx = \int x^{\frac{1}{2}}\,dx = \frac{x^{\frac{3}{2}}}{3/2} + C = \frac{2x^{\frac{3}{2}}}{3} + C$
- $\int \frac{1}{x^5}\,dx = \int x^{-5}\,dx = \frac{x^{-4}}{-4} + C = -\frac{1}{4x^4} + C$
- $\int x^{\pi}\,dx = \frac{x^{\pi+1}}{\pi+1} + C$
- $\int \frac{1}{u}\,du = \ln|u| + C$

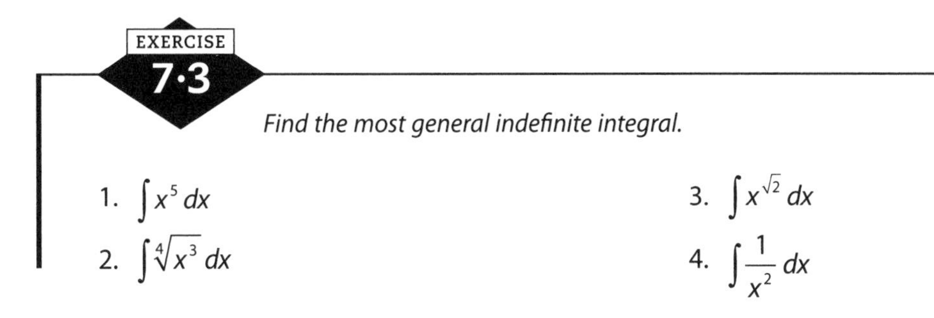
EXERCISE 7·3

*Find the most general indefinite integral.*

1. $\int x^5\,dx$
2. $\int \sqrt[4]{x^3}\,dx$
3. $\int x^{\sqrt{2}}\,dx$
4. $\int \frac{1}{x^2}\,dx$

5. $\int t^{100}\,dt$

6. $\int u^{2\pi}\,du$

7. $\int \frac{1}{\sqrt{x}}\,dx$

8. $\int \frac{x^5}{x^2}\,dx$

9. $\int r^{-1}\,dr$

10. $\int \frac{1}{t}\,dt$

## Integration of exponential functions

The following integral formulas for exponential functions can be derived from the rules for differentiating exponential functions (see Chapter 6) and the chain rule (see Chapter 5):

$$\int e^x\,dx = e^x + C;$$

$$\int e^{kx}\,dx = \frac{1}{k}e^{kx} + C, \text{ for all } k \neq 0;$$

$$\int b^x\,dx = \frac{1}{\ln b}b^x + C, \text{ for all } b > 0,\ b \neq 1; \text{ and}$$

$$\int b^{kx}\,dx = \frac{1}{k\ln b}b^{kx} + C, \text{ for all } b > 0,\ b \neq 1,\ k \neq 0,$$

where $C$ is an arbitrary constant.

- $\int e^u\,du = e^u + C$
- $\int e^{5x}\,dx = \frac{1}{5}e^{5x} + C = \frac{e^{5x}}{5} + C$
- $\int 2^x\,dx = \frac{1}{\ln 2}2^x + C = \frac{2^x}{\ln 2} + C$
- $\int 2^{5x}\,dx = \frac{1}{5\ln 2}2^{5x} + C = \frac{2^{5x}}{5\ln 2} + C$

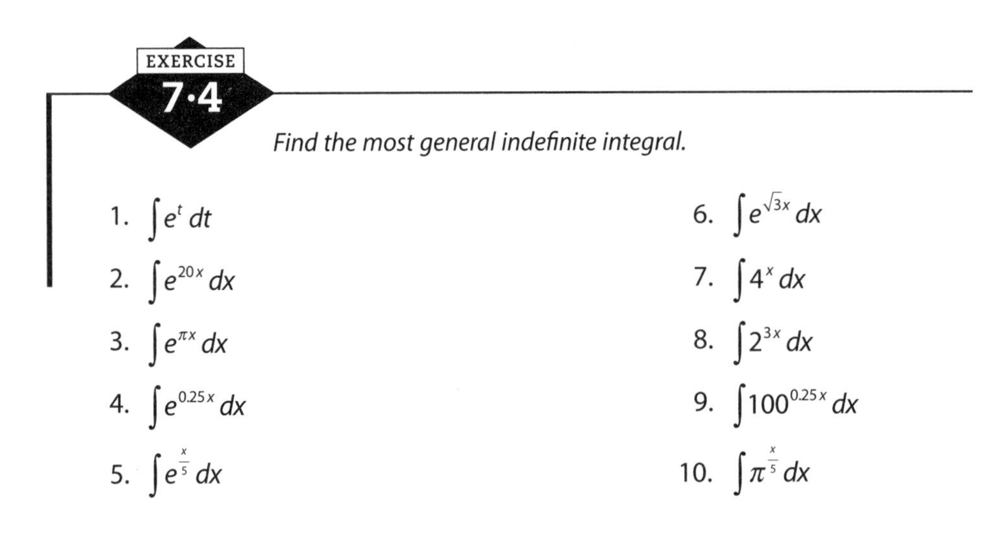

EXERCISE 7·4

*Find the most general indefinite integral.*

1. $\int e^t\,dt$

2. $\int e^{20x}\,dx$

3. $\int e^{\pi x}\,dx$

4. $\int e^{0.25x}\,dx$

5. $\int e^{\frac{x}{5}}\,dx$

6. $\int e^{\sqrt{3}x}\,dx$

7. $\int 4^x\,dx$

8. $\int 2^{3x}\,dx$

9. $\int 100^{0.25x}\,dx$

10. $\int \pi^{\frac{x}{5}}\,dx$

## Integration of derivatives of trigonometric functions

The following integral formulas can be derived from the rules for differentiating the six trigonometric functions (see Chapter 6) and the chain rule (see Chapter 5):

$$\int \sin x\,dx = -\cos x + C;$$

$$\int \sin(kx)\,dx = -\frac{1}{k}\cos(kx) + C, \text{ for all } k \neq 0;$$

$$\int \cos x\,dx = \sin x + C;$$

$$\int \cos(kx)\,dx = \frac{1}{k}\sin(kx) + C, \text{ for all } k \neq 0;$$

$$\int \sec^2 x\,dx = \tan x + C;$$

$$\int \sec^2(kx)\,dx = \frac{1}{k}\tan(kx) + C, \text{ for all } k \neq 0;$$

$$\int \csc^2 x\,dx = -\cot x + C;$$

$$\int \csc^2(kx)\,dx = -\frac{1}{k}\cot(kx) + C, \text{ for all } k \neq 0;$$

$$\int \sec x \tan x\,dx = \sec x + C;$$

$$\int \sec(kx)\tan(kx)\,dx = \frac{1}{k}\sec(kx) + C, \text{ for all } k \neq 0;$$

$$\int \csc x \cot x\,dx = -\csc x + C; \text{ and}$$

$$\int \csc(kx)\cot(kx)\,dx = -\frac{1}{k}\csc(kx) + C, \text{ for all } k \neq 0,$$

where $C$ is an arbitrary constant.

- $\int \sin u\,du = -\cos u + C$
- $\int \cos(10x)\,dx = \frac{1}{10}\sin(10x) + C$
- $\int \sec^2(0.5x)\,dx = \frac{1}{0.5}\tan(0.5x) + C = \frac{\tan(0.5x)}{0.5} + C$
- $\int \csc^2 t\,dt = -\cot t + C$
- $\int \sec\left(\frac{3x}{4}\right)\tan\left(\frac{3x}{4}\right)dx = \int \sec\left(\frac{3}{4}x\right)\tan\left(\frac{3}{4}x\right)dx = \frac{1}{3/4}\sec\left(\frac{3}{4}x\right) + C = \frac{4}{3}\sec\left(\frac{3x}{4}\right) + C$

**Note:** Special techniques are needed to determine the following integrals: $\int \tan x\,dx$, $\int \cot x\,dx$, $\int \sec x\,dx$, and $\int \csc x\,dx$. These integrals can be determined using techniques presented in Chapter 8.

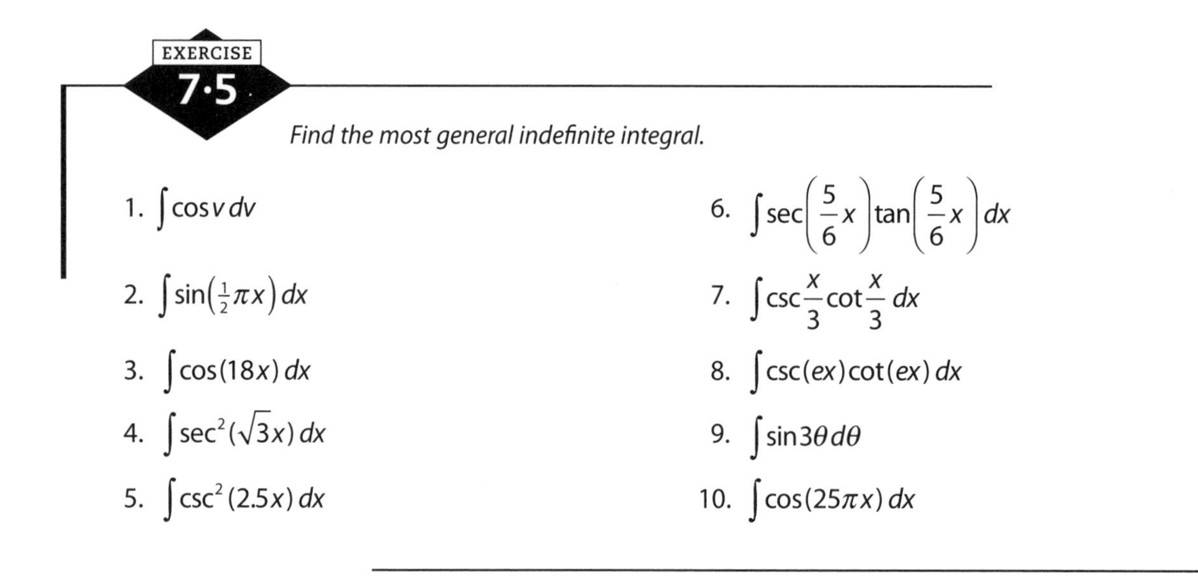

EXERCISE

**7·5**

*Find the most general indefinite integral.*

1. $\int \cos v \, dv$
2. $\int \sin\left(\frac{1}{2}\pi x\right) dx$
3. $\int \cos(18x)\, dx$
4. $\int \sec^2(\sqrt{3}x)\, dx$
5. $\int \csc^2(2.5x)\, dx$
6. $\int \sec\left(\frac{5}{6}x\right)\tan\left(\frac{5}{6}x\right) dx$
7. $\int \csc\frac{x}{3}\cot\frac{x}{3}\, dx$
8. $\int \csc(ex)\cot(ex)\, dx$
9. $\int \sin 3\theta \, d\theta$
10. $\int \cos(25\pi x)\, dx$

# Integration of derivatives of inverse trigonometric functions

The following integral formulas can be derived from the rules for differentiating the six inverse trigonometric functions (see Chapter 6) and the chain rule (see Chapter 5):

$$\int \frac{1}{\sqrt{1-x^2}}\,dx = \sin^{-1} x + C = -\cos^{-1} x + C;$$

$$\int \frac{1}{\sqrt{a^2-x^2}}\,dx = \sin^{-1}\left(\frac{x}{a}\right) + C = -\cos^{-1}\left(\frac{x}{a}\right) + C, \text{ for all } a > 0;$$

$$\int \frac{1}{1+x^2}\,dx = \tan^{-1} x + C = -\cot^{-1} x + C;$$

$$\int \frac{1}{a^2+x^2}\,dx = \frac{1}{a}\tan^{-1}\frac{x}{a} + C = -\frac{1}{a}\cot^{-1}\left(\frac{x}{a}\right) + C, \text{ for all } a > 0;$$

$$\int \frac{1}{|x|\sqrt{x^2-1}}\,dx = \sec^{-1} x + C = -\csc^{-1} x + C; \text{ and}$$

$$\int \frac{1}{|x|\sqrt{x^2-a^2}}\,dx = \frac{1}{a}\sec^{-1}\left(\frac{x}{a}\right) + C = -\frac{1}{a}\csc^{-1}\left(\frac{x}{a}\right) + C, \text{ for all } a > 0,$$

where $C$ is an arbitrary constant.

As you can see from the above formulas, for each integrand that represents a derivative of one of the six inverse trigonometric functions, you have a pair of corresponding antiderivatives from which to choose. This circumstance occurs because the derivatives of the six inverse trigonometric functions fall into three pairs. In each pair, the derivatives differ only in sign. For example, $\frac{d}{dx}(\sin^{-1} x) = \frac{1}{\sqrt{1-x^2}}$ and $\frac{d}{dx}(\cos^{-1} x) = -\frac{1}{\sqrt{1-x^2}}$. When you are integrating an integrand that is the derivative of an inverse trigonometric function, you select only one member from the corresponding pair of antiderivatives. Although mathematically either member is correct, it is customary to select the inverse sine, inverse tangent, and inverse secant functions over the negatives of the inverse cosine, inverse cotangent, and inverse cosecant functions, respectively.

- $\int \frac{du}{\sqrt{1-u^2}} = \int \frac{1}{\sqrt{1-u^2}} du = \sin^{-1} u + C$
- $\int \frac{1}{\sqrt{9-x^2}} dx = \int \frac{1}{\sqrt{3^2-x^2}} dx = \sin^{-1}\left(\frac{x}{3}\right) + C$
- $\int \frac{1}{5+x^2} dx = \int \frac{1}{(\sqrt{5})^2+x^2} dx = \frac{1}{\sqrt{5}} \tan^{-1}\left(\frac{x}{\sqrt{5}}\right) + C$
- $\int \frac{1}{\sqrt{x^2\left(x^2-\frac{36}{25}\right)}} dx = \int \frac{1}{|x|\sqrt{x^2-(\frac{6}{5})^2}} dx = \frac{1}{6/5} \sec^{-1}\left(\frac{x}{6/5}\right) + C = \frac{5}{6} \sec^{-1}\left(\frac{5x}{6}\right) + C$

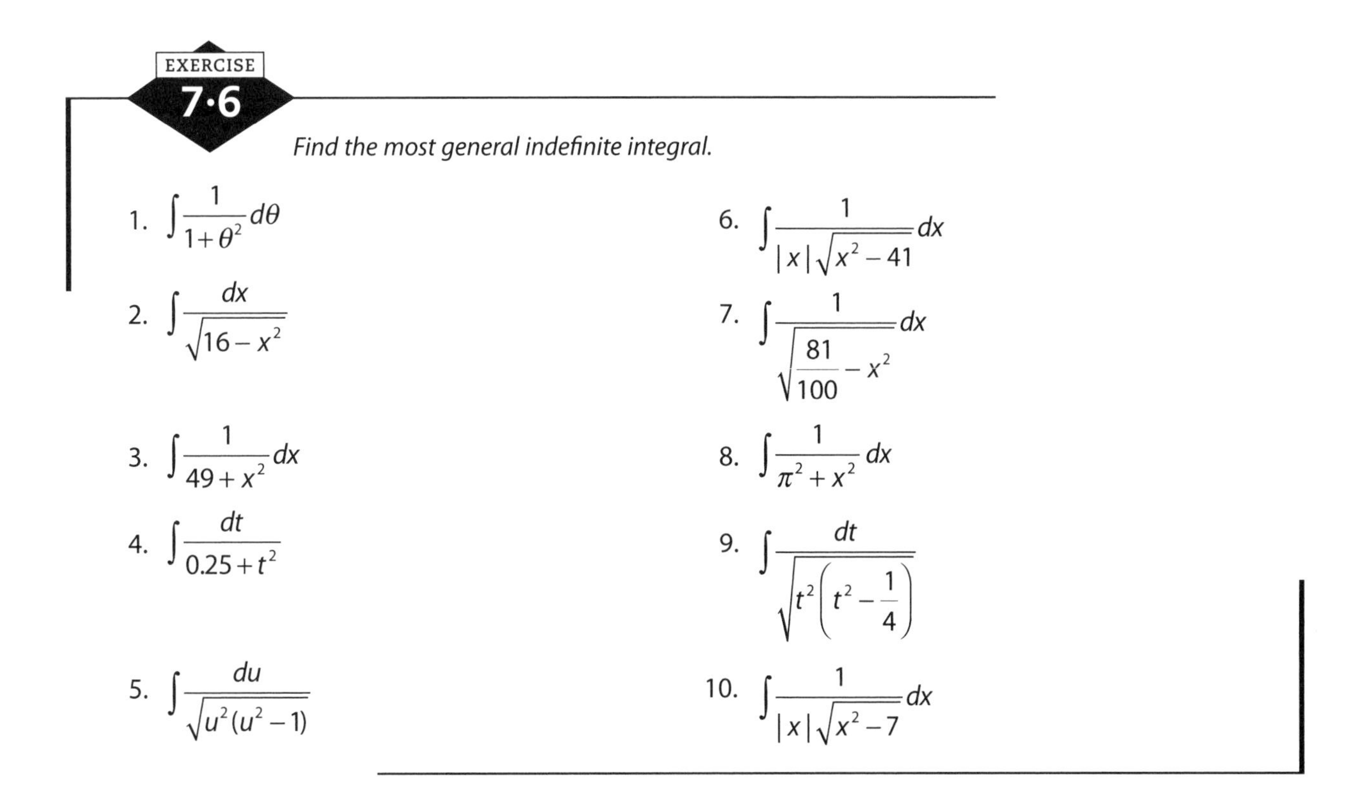

EXERCISE

**7·6**

*Find the most general indefinite integral.*

1. $\int \frac{1}{1+\theta^2} d\theta$
2. $\int \frac{dx}{\sqrt{16-x^2}}$
3. $\int \frac{1}{49+x^2} dx$
4. $\int \frac{dt}{0.25+t^2}$
5. $\int \frac{du}{\sqrt{u^2(u^2-1)}}$
6. $\int \frac{1}{|x|\sqrt{x^2-41}} dx$
7. $\int \frac{1}{\sqrt{\frac{81}{100}-x^2}} dx$
8. $\int \frac{1}{\pi^2+x^2} dx$
9. $\int \frac{dt}{\sqrt{t^2\left(t^2-\frac{1}{4}\right)}}$
10. $\int \frac{1}{|x|\sqrt{x^2-7}} dx$

# Two useful integration rules

Two rules that allow you to integrate a variety of functions are the **constant multiple rule** and the **rule for sums and differences**.

The constant multiple rule states that the integral of a constant times a function is the product of that constant times the integral of the function: Symbolically, you have, if $k$ is any constant, $\int kf(x)dx = k\int f(x)dx$. This rule allows you to factor out constants from an integral, and it applies even when the constant is in the denominator as shown here: $\int \frac{f(x)dx}{k} = \frac{1}{k}\int f(x)dx$, provided $k \neq 0$.

The rule for sums and differences states that the integral of the sum (or difference) of two functions is equal to the sum (or difference) of the integrals of the individual functions. Symbolically, you have $\int[f(x) \pm g(x)]dx = \int f(x)dx \pm \int g(x)dx$.

The example that follows illustrates using the two rules.

$$\int(10x^4 + 6x)dx = \int 10x^4 dx + \int 6x\,dx \text{ by the Rule for Sums and Difference}$$

$$= 10\int x^4 dx + 6\int x\,dx \text{ by the Constant Multiple Rule}$$

$$= 10\left(\frac{x^5}{5}\right) + C_1 + 6\left(\frac{x^2}{2}\right) + C_2 = 2x^5 + 3x^2 + C, \text{ where } C = C_1 + C_2$$

**Note:** It is not necessary to write the two arbitrary constants $C_1$ and $C_2$ separately, as shown in the example above, since their sum $C$ also is an arbitrary constant. When integrating a sum or difference of two or more functions, you can add one constant $C$ to represent the sum of all the arbitrary constants in the solution.

Following are additional examples of applying the constant multiple rule and the rule for sums and differences.

- $\int(5x^4 + 2x^3 - 7x^2 + 12x - 8)dx = \int 5x^4dx + \int 2x^3dx - \int 7x^2dx + \int 12x\,dx - \int 8\,dx =$ $5\int x^4dx + 2\int x^3dx - 7\int x^2dx + 12\int x\,dx - 8\int dx = 5\left(\frac{x^5}{5}\right) + 2\left(\frac{x^4}{4}\right) - 7\left(\frac{x^3}{3}\right) + 12\left(\frac{x^2}{2}\right) -$ $8x + C = x^5 + \frac{1}{2}x^4 - \frac{7}{3}x^3 + 6x^2 - 8x + C$
- $\int[2x - 5\sin(10x)]dx = \int 2x\,dx - \int 5\sin(10x)dx = 2\int x\,dx - 5\int \sin(10x)dx = 2\left(\frac{x^2}{2}\right) -$ $5\left(-\frac{1}{10}\cos 10x\right) + C = x^2 + \frac{1}{2}\cos 10x + C$
- $\int\left(\frac{2}{x} + \frac{3}{x^4}\right)dx = \int\frac{2}{x}dx + \int\frac{3}{x^4}dx = 2\int\frac{1}{x}dx + 3\int x^{-4}dx = 2\ln|x| + 3\left(\frac{x^{-3}}{-3}\right) + C = 2\ln|x| -$ $x^{-3} + C = 2\ln|x| - \frac{1}{x^3} + C$
- $\int\left(\frac{1}{\sqrt{9-x^2}} + \frac{1}{5+x^2}\right)dx = \int\frac{1}{\sqrt{9-x^2}}dx + \int\frac{1}{5+x^2}dx = \int\frac{1}{\sqrt{3^2-x^2}}dx + \int\frac{1}{(\sqrt{5})^2+x^2}dx =$ $\sin^{-1}\left(\frac{x}{3}\right) + \frac{1}{\sqrt{5}}\tan^{-1}\frac{x}{\sqrt{5}} + C$

As it stands, the integral in the next example does not appear to fit any of the basic integration formulas that you've seen thus far. However, simplifying first produces the following:

- $\int\frac{e^{5x} - e^{4x}}{e^x}dx = \int(e^{4x} - e^{3x})\,dx = \int e^{4x}dx - \int e^{3x}dx = \frac{1}{4}e^{4x} - \frac{1}{3}e^{3x} + C$

EXERCISE

**7·7**

*Find the most general indefinite integral.*

1. $\int (3x^4 - 5x^3 - 21x^2 + 36x - 10)\,dx$
2. $\int \left[3x^2 - 4\cos(2x)\right] dx$
3. $\int \left(\frac{8}{t^5} + \frac{5}{t}\right) dt$
4. $\int \left(\frac{1}{\sqrt{25-\theta^2}} + \frac{1}{100+\theta^2}\right) d\theta$
5. $\int \frac{e^{5x} - e^{4x}}{e^{2x}}\,dx$
6. $\int \left(\frac{x^7 + x^4}{x^5}\right) dx$
7. $\int \frac{1}{e^6 + x^2}\,dx$
8. $\int (x^2 + 4)^2\,dx$
9. $\int \left(\frac{7}{\sqrt[3]{t}}\right) dt$
10. $\int \frac{20 + x}{\sqrt{x}}\,dx$

# Basic integration techniques

## Integration by substitution

None of the integration formulas or rules from Chapter 7 fit directly any of the following integrals: $\int (x^2+4)^5 2x\,dx$, $\int e^{x^3}x^2 dx$, or $\int \frac{x^3}{x^4+2}dx$. To find integrals such as these, you can use a method called **integration by substitution** (often called **Integration by $\mu$ – Substitution**). Integration by substitution relies on the chain rule that you used in differentiation. (See Chapter 5 for a discussion of using the chain rule in differentiation.) In integration by substitution you substitute a new variable for a judiciously selected functional expression in the integrand; and then after transforming the original integral, as needed, based on your understanding of the chain rule, you integrate with respect to the new variable. When transforming the integral, the objective is to create an integral that has the form $\int f(g(x))g'(x)\,dx$.

Commonly, the variable $u$ is used as the variable of substitution, as shown in the following examples.

PROBLEM Find $\int (x^2+3)^5 2x\,dx$.

SOLUTION If you let $u = x^2+3$, then $du = 2x\,dx$. When you make these substitutions, the integral takes the form of a power function, which you can integrate as shown below.

$$\int (x^2+3)^5 2x\,dx = \int u^5 du \qquad \text{Substituting } u = x^2+3 \text{ and } du = 2x\,dx$$

$$= \frac{u^6}{6}+C \qquad \text{Integrating with respect to } u$$

$$= \frac{(x+3)^6}{6}+C \qquad \text{Substituting } x^2+3 = u \text{ so that the solution is in terms of the original variable}$$

PROBLEM Find $\int e^{x^3}x^2 dx$.

SOLUTION If you let $u = x^3$, then $du = 3x^2dx$. Since the constant 3 does not appear in the original integrand, you will need to transform the integrand by multiplying the integrand by 1 in the form $\frac{1}{3}\cdot 3$, and then factoring out $\frac{1}{3}$ from the integral, as shown here.

$$\int e^{x^3} x^2 dx = \int e^{x^3} \frac{1}{3} \cdot 3x^2 dx$$ Multiplying by $\frac{1}{3} \cdot 3$

$$= \frac{1}{3}\int e^{x^3} 3x^2 dx = \frac{1}{3}\int e^u du$$ Factoring out $\frac{1}{3}$ and substituting $u = x^3$ and $du = 3x^2 dx$

$$= \frac{1}{3} e^u + C$$ Integrating with respect to $u$

$$= \frac{1}{3} e^{x^3} + C$$ Substituting $x^3 = u$ so that the solution is in terms of the original variable

**Note:** You can use the technique (shown in the above example) of multiplying the integrand by 1 in the form $\frac{1}{k} \cdot k$, and then factoring out $\frac{1}{k}$ from the integral for any nonzero constant $k$; however, a similar technique with variables is not valid. It is incorrect to factor an expression containing the variable from an integral.

PROBLEM Find $\int \frac{x^3}{x^4 + 2} dx$.

SOLUTION If you let $u = x^4 + 2$, then $du = 4x^3 dx$. Since the constant 4 does not appear in the original integrand, you will need to transform the integral by multiplying the integrand by 1 in the form $\frac{1}{4} \cdot 4$, and then factoring out $\frac{1}{4}$ from the integral, as shown here.

$$\int \frac{x^3}{x^4+2} dx = \int \frac{1}{4} \cdot \frac{4x^3}{x^4+2} dx$$ Multiplying by $\frac{1}{4} \cdot 4$

$$= \frac{1}{4}\int \frac{4x^3}{x^4+2} dx = \frac{1}{4}\int \frac{du}{u}$$ Factoring out $\frac{1}{4}$ and substituting $u = x^4 + 2$ and $du = 4x^3 dx$

$$= \frac{1}{4}\ln|u| + C$$ Integrating with respect to $u$

$$= \frac{1}{4}\ln(x^4 + 2) + C$$ Substituting $x^4 + 2 = u$ so that the solution is in terms of the original variable

Becoming adept at choosing $u$-substitutions takes practice. You should memorize the basic integration formulas presented in Chapter 7 to facilitate the process. Here are some general guidelines: Substitute $u$ for

- an expression in parentheses
- the exponent in an exponential expression
- the denominator of a fraction or
- an expression under a radical sign (except when the integrand has the form of the derivative of an inverse sine or secant function)

EXERCISE
**8·1**

*Use integration by substitution to find the most general indefinite integral.*

1. $\int 3(x^3-5)^4 x^2 dx$
2. $\int e^{x^4} x^3 dx$
3. $\int \frac{t}{t^2+7} dt$
4. $\int (x^5-3x)^{\frac{1}{4}}(5x^4-3)dx$
5. $\int \frac{x^3-2x}{(x^4-4x^2+5)^4} dx$
6. $\int \frac{x^3-2x}{x^4-4x^2+5} dx$
7. $\int x\cos(3x^2+1)dx$
8. $\int \frac{3\cos^2\sqrt{x}(\sin\sqrt{x})}{\sqrt{x}} dx$
9. $\int \frac{e^{2x}}{1+e^{4x}} dx$
10. $\int 6t^2 e^{t^3-2} dt$

# Integration by parts

**Integration by parts** is a powerful technique for integrating certain complicated integrals such as $\int x\sin 3x\,dx$, $\int x^5 \ln x\,dx$, and $\int x^2 e^{-x} dx$ that do not lend themselves to basic integration formulas or to the technique of integration by substitution. If $u$ and $v$ are differentiable functions, then the equation for integration by parts is given by

$$\int u\,dv = u\cdot v - \int v\cdot du$$

The given integral is $\int u\,dv$, which has two "parts": $u$ and $dv$. The goal of integration by parts is to wisely select these two parts so that the resulting integral on the right, $\int v\cdot du$, is easier to integrate than the original integral on the left, $\int u\,dv$. To see how the formula works, consider the following examples.

PROBLEM Find $\int x\sin 3x\,dx$.

SOLUTION Let $u = x$ and $dv = \sin 3x\,dx$.

Then $du = dx$ and $v = \int \sin 3x\,dx = -\frac{1}{3}\cos 3x$.

**Note:** The constant of integration is added at the end of the process.

Now, using the integration by parts equation, you have

$$\int u\,dv = u\cdot v - \int v\cdot du$$

$$\int x\sin 3x\,dx = (x)\left(-\frac{1}{3}\cos 3x\right) - \int\left(-\frac{1}{3}\cos 3x\right)\cdot dx$$

$$= -\frac{1}{3}x\cos 3x + \frac{1}{3}\int \cos 3x\,dx$$

$$= -\frac{1}{3}x\cos 3x + \frac{1}{3}\cdot\frac{1}{3}\sin 3x$$
$$= -\frac{1}{3}x\cos 3x + \frac{1}{9}\sin 3x + C.$$

PROBLEM Find $\int x^5 \ln x\, dx$.

SOLUTION Let $u = \ln x$ and $dv = x^5 dx$.

Then $du = \frac{1}{x}dx$ and $v = \int x^5 dx = \frac{x^6}{6}$.

Now, using the integration by parts equation, you have

$$\int u\, dv = u\cdot v - \int v\cdot du$$
$$\int x^5 \ln x\, dx = (\ln x)\left(\frac{x^6}{6}\right) - \int \frac{x^6}{6}\cdot\frac{1}{x}dx$$
$$= \frac{x^6 \ln x}{6} - \frac{1}{6}\int x^5 dx$$
$$= \frac{x^6 \ln x}{6} - \frac{1}{6}\cdot\frac{x^6}{6}$$
$$= \frac{x^6 \ln x}{6} - \frac{x^6}{36} + C.$$

Sometimes, you might need to apply integration by parts more than once as shown in the following example.

PROBLEM Find $\int x^2 e^{-x} dx$.

SOLUTION Let $u = x^2$ and $dv = e^{-x}dx$.

Then $du = 2x\, dx$ and $v = \int e^{-x} dx = -e^{-x}$.

Now, using the integration by parts equation, you have

$$\int u\, dv = u\cdot v - \int v\cdot du$$
$$\int x^2 e^{-x} dx = (x^2)(-e^{-x}) - \int -e^{-x}\cdot 2x\, dx$$
$$= -x^2 e^{-x} + \int 2xe^{-x} dx$$

As you can see, the integral on the right does not fit a basic integration formula. To integrate that integral, you can apply integration by parts for a second time.

This time, let $u = 2x$ and $dv = e^{-x}dx$.

Then $du = 2\, dx$ and $v = \int e^{-x}dx = -e^{-x}$.

Now, using the integration by parts equation, you have

$$\int u\, dv = u\cdot v - \int v\cdot du$$
$$\int 2xe^{-x} dx = (2x)(-e^{-x}) - \int -e^{-x}\cdot 2\, dx$$
$$= -2xe^{-x} + 2\int e^{-x} dx$$
$$= -2xe^{-x} - 2e^{-x}$$

Combining these two results, you have

$$\int x^2e^{-x}dx = -x^2e^{-x} - 2xe^{-x} - 2e^{-x} + C.$$

Here are some general guidelines to follow for integration by parts.

1. Try letting $dv$ be the most complicated part of the integrand that you recognize as integrable.
2. Always include the differential as part of $dv$.
3. Try letting $u$ be a portion of the integrand whose derivative is simpler than $u$.
4. For integrals that consist of a single factor times the differential, let $dv$ be the differential.
5. Be prepared to apply integration by parts more than once within the same problem.

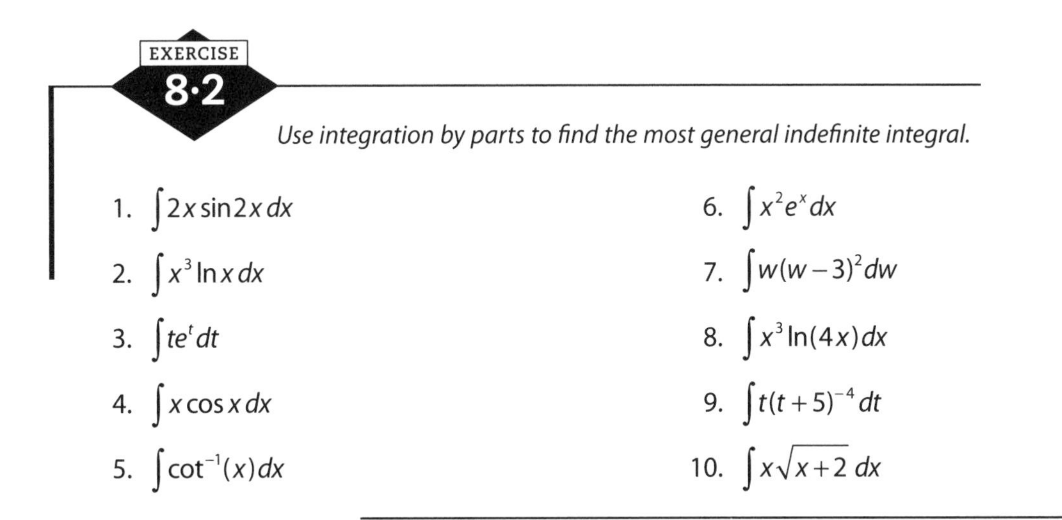

EXERCISE 8·2

*Use integration by parts to find the most general indefinite integral.*

1. $\int 2x \sin 2x\, dx$
2. $\int x^3 \ln x\, dx$
3. $\int te^t dt$
4. $\int x \cos x\, dx$
5. $\int \cot^{-1}(x)dx$
6. $\int x^2e^x dx$
7. $\int w(w-3)^2 dw$
8. $\int x^3 \ln(4x)dx$
9. $\int t(t+5)^{-4} dt$
10. $\int x\sqrt{x+2}\, dx$

# Integration by using tables of integral formulas

Another technique of integration is to integrate by using tables of integral formulas. A table of 67 common integral formulas is provided in Appendix C for your convenience. The following is some helpful information about tables of integrals in general:

1. Letters at the beginning of the alphabet (e.g., $a$, $b$, $c$, and $d$) represent constants.
2. The letter $n$ is often used to represent a constant exponent (e.g., $x^n$).
3. The letter $k$ is often used to represent a constant in an exponential expression (e.g., $e^{kx}$).
4. If the integrand contains a fraction, the differential might be in the numerator of the fraction.
5. The constant of integration might be omitted.
6. The natural logarithm might be written as $\log(x)$ instead of $\ln x$.

To integrate using a table of integral formulas, you simply look through the table until you find an integral formula in which the integrand exactly matches the form of the integrand of the integral you want to integrate. Sometimes, the task of locating such an integral formula is straightforward as in the following examples.

PROBLEM Find $\int \frac{1}{1+e^x} dx$.

SOLUTION This integral matches Formula 38. Therefore,

$$\int \frac{1}{1+e^x} dx = x - \ln(1+e^x) + C$$

PROBLEM Find $\int \tan u \, du$.

SOLUTION This integral matches Formula 13. Therefore,

$$\int \tan u \, du = -\ln|\cos u| + C$$

PROBLEM Find $\int \ln t \, dt$.

SOLUTION This integral matches Formula 40. Therefore,

$$\int \ln t \, dt = t \ln t - t + C$$

In some cases, you might need to substitute values for constants that appear in the formula as shown in this example.

PROBLEM Find $\int \frac{1}{x\sqrt{3x+5}} dx$.

SOLUTION This integral matches Formula 55 with $a = 3$ and $b = 5$. Therefore, you have

$$\int \frac{1}{x\sqrt{ax+b}} dx = \frac{1}{\sqrt{b}} \ln\left|\frac{\sqrt{ax+b}-\sqrt{b}}{\sqrt{ax+b}+\sqrt{b}}\right| + C$$

$$\int \frac{1}{x\sqrt{3x+5}} dx = \frac{1}{\sqrt{5}} \ln\left|\frac{\sqrt{3x+5}-\sqrt{5}}{\sqrt{3x+5}+\sqrt{5}}\right| + C$$

Sometimes, the appropriate integration formula might be difficult to find. Before giving up, experiment with the following techniques to try to transform the given integral into one for which you can use a table of integral formulas.

1. Expand expressions that are raised to a power.
2. Rewrite expressions that are raised to a negative power as equivalent expressions raised to a positive power.
3. Factor out extraneous constants from the integral.
4. Separate a numerator that has more than one term into separate algebraic fractions.
5. Write an improper algebraic fraction as a quotient plus remainder over denominator.
6. Complete the square for a quadratic expression.

If this line of attack fails, you might have to concede that the integral cannot be integrated using basic methods.

**Note:** Certain graphing calculators (e.g., the *TI-92*) and some software programs (e.g., *Derive*, *Maple*, and *Mathematica*) are capable of producing symbolic results of integration. However, it is not unusual for the results to differ from what you obtain through traditional means. Furthermore, you might find that the symbolic integration tool is unable to find the antiderivative for the integrand. Nevertheless, a symbolic integration tool can be useful for performing integration of complicated integrals. Understanding the process as shown in this chapter will greatly benefit you when using such a utility.

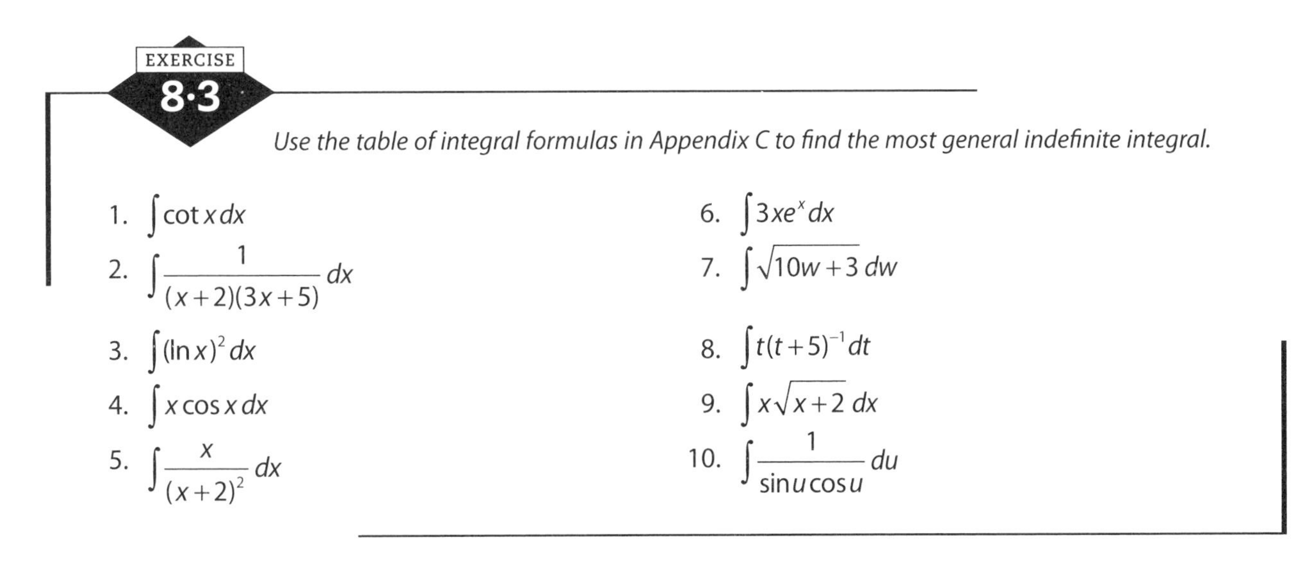

EXERCISE 8·3

*Use the table of integral formulas in Appendix C to find the most general indefinite integral.*

1. $\int \cot x \, dx$
2. $\int \frac{1}{(x+2)(3x+5)} dx$
3. $\int (\ln x)^2 dx$
4. $\int x \cos x \, dx$
5. $\int \frac{x}{(x+2)^2} dx$
6. $\int 3xe^x dx$
7. $\int \sqrt{10w+3} \, dw$
8. $\int t(t+5)^{-1} dt$
9. $\int x\sqrt{x+2} \, dx$
10. $\int \frac{1}{\sin u \cos u} du$

# The definite integral

## Definition of the definite integral and the First Fundamental Theorem of Calculus

If a function $f$ is defined on the closed interval $[a, b]$, then the **definite integral** of $f$ from $a$ to $b$ is defined as a limiting sum given by: $\int_a^b f(x)\,dx = \lim_{\max \Delta x_i \to 0} \sum_{i=1}^{n} f(c_i)\Delta x_i$, where $[a, b]$ is divided into $n$ subintervals (not necessarily equal), $c_i$ is a point in the $i$th subinterval $[x_{i-1}, x_i]$, and $\Delta x_i = x_i - x_{i-1}$, provided this limit exists.

The limiting sum, $\sum_{i=1}^{n} f(c_i)\Delta x_i$, in the definition of the definite integral is called a **Riemann sum**. This sum is a numerical result.

Fortunately, the following theorem means that for continuous functions there is a powerful method to evaluate definite integrals rather than the use of Riemann sums.

The **First Fundamental Theorem of Calculus**: If $f$ is continuous on the closed interval $[a, b]$ and $F$ is an antiderivative of $f$ on $[a, b]$, then the evaluation of the **definite integral** $\int_a^b f(x)\,dx$ is given by $\int_a^b f(x)\,dx = F(b) - F(a)$.

This theorem means that you can evaluate the definite integral, $\int_a^b f(x)\,dx$, through a four-step process:

1. Determine $F$ an antiderivative of $f$.
2. Evaluate $F(b)$.
3. Evaluate $F(a)$.
4. Calculate $F(b) - F(a)$.

**Note:** The constant of integration is subtracted out when a definite integral is evaluated. Therefore, you can omit it from the calculations.

The following notations are used when applying the Fundamental Theorem of Calculus to evaluate the definite integral,

$$\int_a^b f(x)\,dx = F(b) - F(a) = F(x)\Big|_a^b = [F(x)]_a^b = [F(x)]_a^b = [F(x)]_{x=a}^{x=b}$$

**Note:** Hereafter, the symbol $\approx$ will be used to mean "approximately equal to."

PROBLEM Evaluate $\int_1^4 15x^2\,dx$.

SOLUTION $\int_1^4 15x^2\,dx = 5x^3\Big|_1^4 = 5(4)^3 - 5(1)^3 = 320 - 5 = 315$

PROBLEM Evaluate $\int_0^{\frac{\pi}{4}} \sec^2\theta\, d\theta$.

SOLUTION $\int_0^{\frac{\pi}{4}} \sec^2\theta\, d\theta = \tan\theta\Big|_0^{\frac{\pi}{4}} = \tan\left(\frac{\pi}{4}\right) - \tan(0) = 1 - 0 = 1$

PROBLEM Evaluate $\int_{-2}^{2} (10x^4 + 6x)\,dx$.

SOLUTION $\int_{-2}^{2} (10x^4 + 6x)\,dx = (2x^5 + 3x^2)\Big|_{-2}^{2} = (2\cdot 2^5 + 3\cdot 2^2) - (2\cdot(-2)^5 + 3\cdot(-2)^2) =$

$76 + 52 = 128$

PROBLEM Evaluate $\int_{-1}^{2} e^{x^3} x^2\,dx$.

SOLUTION $\int_{-1}^{2} e^{x^3} x^2\,dx = \frac{1}{3}\int_{-1}^{2} e^{x^3} 3x^2\,dx = \frac{1}{3}e^{x^3}\Big|_{-1}^{2} = \frac{1}{3}e^{(2)^3} - \frac{1}{3}e^{(-1)^3} = \frac{1}{3}e^{8} - \frac{1}{3}e^{-1} =$

$993.6526... - .1226... \approx 993.53$ (Note: Avoid rounding until the final calculation.)

PROBLEM Evaluate $\int_{3}^{10} \ln t\,dt$.

SOLUTION $\int_{3}^{10} \ln t\,dt = (t\ln t - t)\Big|_{3}^{10} = (10\ln 10 - 10) - (3\ln 3 - 3) = 13.0258... - 0.2958... \approx 12.73$

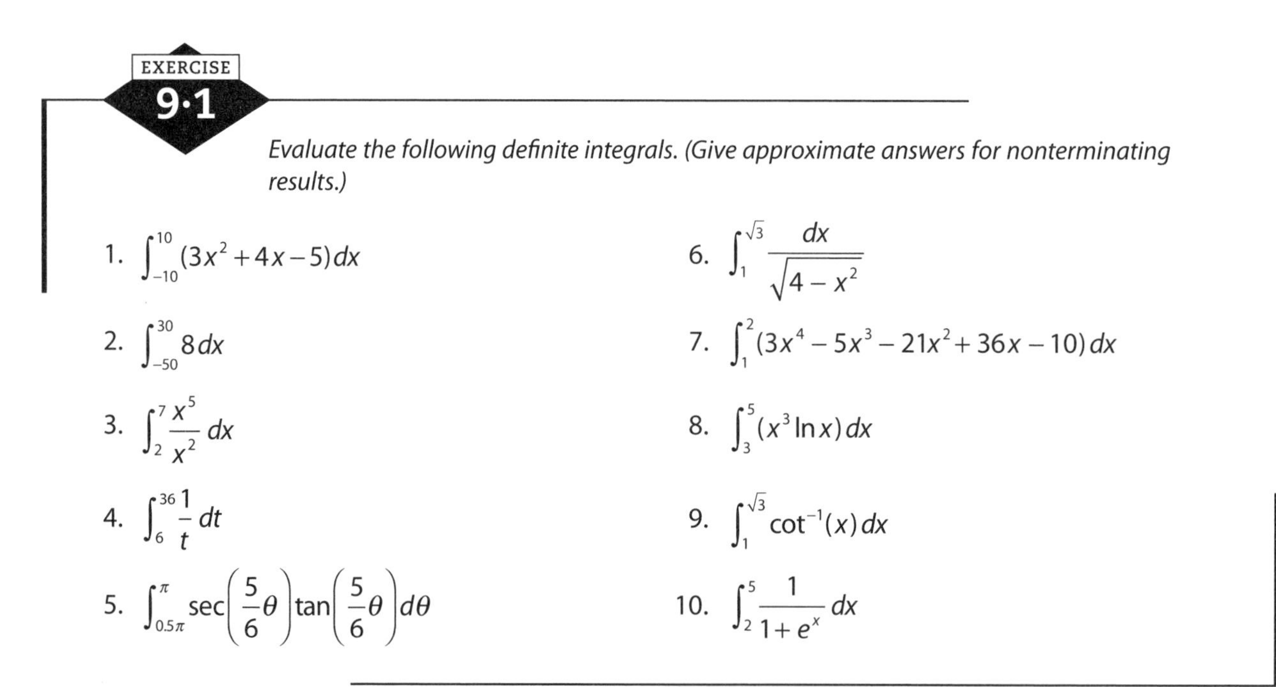

EXERCISE 9·1

*Evaluate the following definite integrals. (Give approximate answers for nonterminating results.)*

1. $\int_{-10}^{10} (3x^2 + 4x - 5)\,dx$
2. $\int_{-50}^{30} 8\,dx$
3. $\int_{2}^{7} \frac{x^5}{x^2}\,dx$
4. $\int_{6}^{36} \frac{1}{t}\,dt$
5. $\int_{0.5\pi}^{\pi} \sec\left(\frac{5}{6}\theta\right)\tan\left(\frac{5}{6}\theta\right)d\theta$
6. $\int_{1}^{\sqrt{3}} \frac{dx}{\sqrt{4 - x^2}}$
7. $\int_{1}^{2} (3x^4 - 5x^3 - 21x^2 + 36x - 10)\,dx$
8. $\int_{3}^{5} (x^3 \ln x)\,dx$
9. $\int_{1}^{\sqrt{3}} \cot^{-1}(x)\,dx$
10. $\int_{2}^{5} \frac{1}{1 + e^x}\,dx$

# Useful properties of the definite integral

The definite integral has the following useful properties.

1. If $f$ is defined at $x = a$, then $\int_a^a f(x)\,dx = 0$.
2. If $f$ is integrable on $[a, b]$, then $\int_a^b f(x)\,dx = -\int_b^a f(x)\,dx$.

3. If $f$ is integrable on $[a, b]$, $[a, c]$, and $[c, b]$, then $\int_a^b f(x)dx = \int_a^c f(x)dx + \int_c^b f(x)dx$.

4. If $f$ is integrable on $[a, b]$ and $k$ is a constant, then $\int_a^b kf(x)dx = k\int_a^b f(x)dx$.

5. If $f$ and $g$ are integrable on $[a, b]$, then $\int_a^b [f(x) \pm g(x)]dx = \int_a^b f(x)dx \pm \int_a^b g(x)dx$.

6. If $f$ is integrable and nonnegative on $[a, b]$, then $\int_a^b f(x)dx \geq 0$.

7. If $f$ and $g$ are integrable on $[a, b]$ and if $f(x) \leq g(x)$ for every $x$ in $[a, b]$, then $\int_a^b f(x)dx \leq \int_a^b g(x)dx$.

PROBLEM Given $\int_0^4 f(x)dx = 25$ and $\int_4^9 f(x)dx = 40$, find

(a) $\int_4^0 f(x)dx$

(b) $\int_0^9 f(x)dx$

(c) $\int_4^4 f(x)dx$

(d) $\int_0^4 2f(x)dx$

SOLUTION (a) By Property 2, $\int_4^0 f(x)dx = -\int_0^4 f(x)dx = -25$.

(b) By Property 3, $\int_0^9 f(x)dx = \int_0^4 f(x)dx + \int_4^9 f(x)dx = 25 + 40 = 65$.

(c) By Property 1, $\int_4^4 f(x)dx = 0$.

(d) By Property 4, $\int_0^4 2f(x)dx = 2\int_0^4 f(x)dx = 2(25) = 50$.

PROBLEM Given $\int_{-5}^5 f(x)dx = 6$ and $\int_{-5}^5 g(x)dx = -4$, find

(a) $\int_{-5}^5 [f(x)+g(x)]\,dx$

(b) $\int_{-5}^5 [f(x)-g(x)]\,dx$

SOLUTION (a) By Property 5, $\int_{-5}^5 [f(x)+g(x)]\,dx = \int_{-5}^5 f(x)dx + \int_{-5}^5 g(x)dx = 6 + (-4) = 2$.

(b) By Property 5, $\int_{-5}^5 [f(x)-g(x)]\,dx = \int_{-5}^5 f(x)dx - \int_{-5}^5 g(x)dx = 6 - (-4) = 10$.

EXERCISE

## 9·2

*For problems 1–6, evaluate the definite integral, given* $\int_{-2}^0 f(x)dx = 12$ *and* $\int_0^2 f(x)dx = 15$. *Justify your answer.*

1. $\int_2^2 f(x)dx$

2. $\int_0^{-2} f(x)dx$

3. $\int_1^1 f(x)dx$

4. $\int_{-2}^2 f(x)dx$

5. $\int_{-2}^0 5f(x)dx$

6. $\int_2^{-2} 10f(x)dx$

*For problems 7–10, evaluate the definite integral, given* $\int_1^5 f(x)dx = -8$ *and* $\int_1^5 g(x)dx = 22$. *Justify your answer.*

7. $\int_1^5 [f(x)+g(x)]\,dx$

8. $\int_1^5 [f(x)-g(x)]\,dx$

9. $\int_1^5 \frac{1}{2}f(x)\,dx$

10. $\int_1^5 2g(x)\,dx + \int_1^5 3f(x)\,dx$

# Second Fundamental Theorem of Calculus

The **Second Fundamental Theorem of Calculus** states that if $f$ is continuous on the closed interval $[a, b]$, then the function $F$ defined by

$$F(x) = \int_a^x f(t)\,dt, \text{ where } x \text{ is in } [a, b]$$

is differentiable on $[a, b]$ and is an antiderivative of $f$; that is to say, for every $x$ in $[a, b]$,

$$F'(x) = \frac{d}{dx}\left[\int_a^x f(t)\,dt\right] = f(x)$$

**Note:** To avoid confusion, since the variable $x$ is used as the upper limit in the integral, $\int_a^x f(t)\,dt$, the variable $t$ is used as the variable of integration.

The following examples illustrate this useful theorem.

- $\frac{d}{dx}\left[\int_0^x \sin(t)\,dt\right] = \sin(x)$
- $\frac{d}{dx}\left[\int_1^x t^3\,dt\right] = x^3$
- $\frac{d}{dx}\left[\int_{-1}^x \sqrt{t+2}\,dt\right] = \sqrt{x+2}$

PROBLEM Find $F'(x)$, when $F(x) = \int_3^x \frac{1}{t}\,dt$.

SOLUTION $F'(x) = \frac{d}{dx}\left[\int_3^x \frac{1}{t}\,dt\right] = \frac{1}{x}$

The next example illustrates applying the chain rule in conjunction with the Second Fundamental Theorem of Calculus.

$$\frac{d}{dx}\left[\int_0^{3x^2} \sin(t)\,dt\right] = \sin(3x^2)\cdot\frac{d}{dx}(3x^2) = \sin(3x^2)\cdot 6x = 6x\sin(3x^2)$$

The Second Fundamental Theorem of Calculus guarantees that if a function is continuous, then it has an antiderivative. However, the antiderivative might not be readily obtainable.

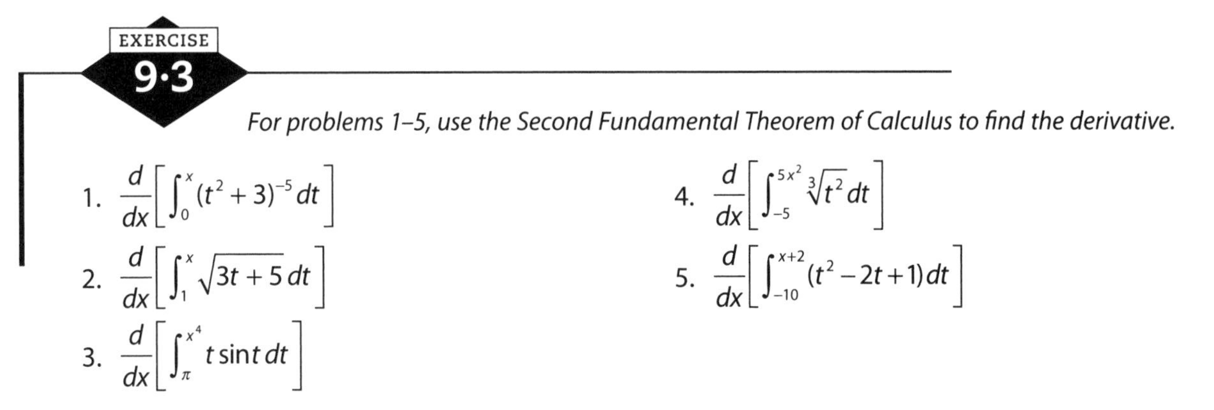

EXERCISE

**9·3**

*For problems 1–5, use the Second Fundamental Theorem of Calculus to find the derivative.*

1. $\frac{d}{dx}\left[\int_0^x (t^2+3)^{-5}\,dt\right]$
2. $\frac{d}{dx}\left[\int_1^x \sqrt{3t+5}\,dt\right]$
3. $\frac{d}{dx}\left[\int_\pi^{x^4} t\sin t\,dt\right]$
4. $\frac{d}{dx}\left[\int_{-5}^{5x^2} \sqrt[3]{t^2}\,dt\right]$
5. $\frac{d}{dx}\left[\int_{-10}^{x+2} (t^2-2t+1)\,dt\right]$

*For problems 6–10, use the Second Fundamental Theorem of Calculus to find* $F'(x)$.

6. $F(x)=\int_0^x \sin(3t)\,dt$
7. $F(x)=\int_5^{4x} \frac{1}{t+1}\,dt$
8. $F(x)=\int_0^{\sin x} 6t^2\,dt$
9. $F(x)=\int_{-3}^{\sqrt{x}} 2t^4\,dt$
10. $F(x)=\int_{-8}^{2x+1} (3t-7)\,dt$

# Mean Value Theorem for Integrals

**The Mean Value Theorem for Integrals** states that if $f$ is continuous on the closed interval $[a, b]$, then there exists a number $c$ in $[a, b]$ such that

$$\int_a^b f(x)\,dx = f(c)(b-a)$$

This theorem guarantees that the number $c$ exists in $[a, b]$, but notice that the theorem does not specify the value of $c$. Many of the problems associated with this concept involve finding values of $c$. On the other hand, in some cases, it may be sufficient just to know that at least one such number in $[a, b]$ exists.

PROBLEM Find the value of $c$ guaranteed by the Mean Value Theorem for Integrals for the function defined by $f(x)=x^2+2$ and the interval $[0, 3]$.

SOLUTION By the Mean Value Theorem for Integrals, you have

$$\int_a^b f(x)\,dx = f(c)(b-a)$$

$$\int_0^3 (x^2+2)\,dx = (c^2+2)(3-0)$$

$$\left(\frac{x^3}{3}+2x\right)\Bigg|_0^3 = 3c^2+6$$

$$\left(\frac{(3)^3}{3}+2(3)\right)-\left(\frac{(0)^3}{3}+2(0)\right) = 3c^2+6$$

$$15-0 = 3c^2+6$$

$$9 = 3c^2$$

$$3 = c^2$$

$$\pm\sqrt{3} = c$$

Of these two possible values for $c$, only the value $\sqrt{3}$ lies in [0, 3], so $c = \sqrt{3}$ is the value guaranteed by the Mean Value Theorem for Integrals.

If $f$ is integrable on the closed interval $[a, b]$, the **average value** of $f$ is

$$\frac{1}{b-a}\int_a^b f(x)\,dx$$

In other words, the value of $f(c)$ given in the Mean Value Theorem for Integrals is the average value of $f$ on the interval $[a, b]$.

PROBLEM Find the average value of $f(x) = x^2 + 2$ on the interval [0, 3].

SOLUTION The average value is given by

$$\frac{1}{b-a}\int_a^b f(x)\,dx = \frac{1}{3}\int_0^3 (x^2+2)\,dx$$

$$= \frac{1}{3}\left(\frac{x^3}{3}+2x\right)\Bigg|_0^3$$

$$= \frac{1}{3}\left[\left(\frac{(3)^3}{3}+2(3)\right)-\left(\frac{(0)^3}{3}+2(0)\right)\right] = \frac{1}{3}[(15)-(0)] = \frac{1}{3}[15] = 5$$

*In problems 1–5, find the value of c guaranteed by the Mean Value Theorem for Integrals for the given function over the indicated interval.*

1. $f(x) = 2x + 6$ and the interval [–1, 1]
2. $f(x) = 2 - 5\sqrt{x}$ and the interval [0, 4]
3. $f(x) = \dfrac{4}{x^3}$ and the interval [1, 4]
4. $f(x) = \sin x$ and the interval $[0, \pi]$
5. $f(x) = \dfrac{1}{x}$ and the interval [1, 3]

*For problems 6–10, find the average value of the given function over the indicated interval.*

6. $f(x) = x^2$ and the interval [–2, 2]
7. $f(x) = \dfrac{1}{x}$ and the interval [1, 3]
8. $f(x) = \cos x$ and the interval $\left[-\dfrac{\pi}{2}, \dfrac{\pi}{2}\right]$
9. $f(x) = \dfrac{9}{2}\sqrt{x}$ and the interval [1, 4]
10. $f(x) = e^x$ and the interval [0, 1]

# APPLICATIONS OF THE DERIVATIVE AND THE DEFINITE INTEGRAL

This culminating Part IV highlights some applications of differential and integral calculus. Of course, there are many other applications since calculus is used in virtually every branch of the physical sciences and also in engineering, computer science, statistics, economics, business, medicine, and in numerous other real-world venues. Nonetheless, this material is designed to give you an appreciation of the versatility and power of calculus and why it is an important and valuable mathematical tool.

# Applications of the derivative

## Slope of the tangent line at a point

If $f'(a)$ exists, then the **slope of the tangent line** to the graph of the function $f$ at the point $P(a, f(a))$ is the line through $P$ that has slope $m = f'(a)$.

PROBLEM Find the equation of the tangent line to the parabola $y = f(x) = x^2 + 1$ at the point (2, 5).

SOLUTION $f'(x) = 2x$, so $m = f'(2) = 4$. Now, since the point (2, 5) is on the tangent, the equation of the desired line is $y - y_1 = m(x - x_1)$, or in this case $y - 5 = 4(x - 2)$, which gives $y = 4(x - 2) + 5$ and finally $y = 4x - 3$.

PROBLEM Find the equation of the tangent line to $y = g(x) = e^{3x}$ at the $y$-intercept.

SOLUTION Since $g'(x) = 3e^{3x}$, the solution is given by the equation $y = g'(0)(x - 0) + 1$ since the $y$-intercept occurs at (0, 1). Hence the required equation is $y = 3x + 1$.

PROBLEM Find all the points on the curve $y = f(x) = \sqrt{x^4 + x^2}$ where the tangent line is horizontal.

SOLUTION The tangent line will be horizontal at any point $x$ where it has zero slope; that is, when $f'(x) = 0$. Inspection of $f'(x) = \dfrac{(2x^3 + x)}{(x^4 + x^2)^{\frac{1}{2}}}$ reveals that $x = 0$ is the only value that could possibly make the derivative zero; but at this value, $f'$ is undefined. Thus, there is no solution to this problem.

PROBLEM Find all the points on the curve $y = f(x) = e^{x^2 - \cos x}$ where the tangent line is horizontal.

SOLUTION $f'(x) = (2x + \sin x)e^{x^2 - \cos x}$. So if you set $f'(x) = 0$ and solve, you have $e^{x^2 - \cos x} = 0$ or $(2x + \sin x) = 0$. The first expression is never equal to 0. Therefore, you solve $(2x + \sin x) = 0$ or $2x = -\sin x$ which, by inspection, is true only when is $x = 0$. Thus, $(0, f(0)) = (0, e^{-\cos 0}) = \left(0, \dfrac{1}{e}\right)$ is the solution.

EXERCISE
**10·1**

*Solve the following.*

1. Find the slope of the tangent line to $f(x) = x^3 + e^x + \sin(x)$ at $x = -1$.
2. Find the slope of the tangent line to $f(x) = \ln(x-1) + x^2$ at $x = 2$.
3. Find the equation of the line tangent to the curve $y = 2x^2 + 4x$ at $(-2, 0)$.
4. Find the slope of the line tangent to the curve at $(x, f(x))$ for $f(x) = x^3 - 6x^2 + 9x - 2$.
5. Find all points on the curve $f(x) = x^2 - 4\sqrt{x} + 1$ where the tangent line is horizontal.
6. Find all points on the curve $f(x) = x^5 - 5x^3 - 20x + 7$ where the tangent line is horizontal.
7. Find the equation of the line tangent to the curve $x^2 + 3xy + y^2 = 5$ at (1, 1).
8. Find the equation of the tangent line to the curve $y = 2x^2 + 3$ that is parallel to the line $y = 8x + 3$.
9. Find the equation of the line tangent to the curve $y = 4 - x^2$ at (1, 2).
10. Find the equation of the line tangent to the curve $f(x) = \dfrac{1 - \sin x}{x + 1}$ at $x = (0, 1)$.

# Instantaneous rate of change

If $f'(t)$ exists, then the (instantaneous) **rate of change** of $f$ at $t$ is $f'(t)$. For example, if $s(t)$ is the position function of a moving object at time $t$, then the **velocity** $v$, the instantaneous rate of change, of the object at time $t$ is $s'(t) = v(t)$. (This is yet another interpretation of the derivative.) Additionally, the **acceleration** $a$ of the object at time $t$ is $s''(t) = v'(t) = a(t)$.

The sign of the velocity function indicates the direction in which the object is moving. When $v(t) > 0$, the object is moving to the right, and when $v(t) < 0$, the object is moving to the left. Furthermore, as logic would dictate, at the instant that a moving object changes direction, $v(t) = 0$ (since the object must stop in order to change direction).

**Speed** is defined as the absolute value of the velocity. That definition is the reason the main dial on an automobile is called a speedometer. It gives you the speed, but not the direction of travel.

PROBLEM Discuss the motion of a particle that moves along a horizontal line so that the position $s$ of the particle on the horizontal line is a function of time $t$ according to the equation $s(t) = t^3 - 2t^2 + t$.

SOLUTION Differentiating the position function with respect to time gives the velocity function, $s'(t) = v(t) = 3t^2 - 4t + 1$. A quick analysis of this quadratic function indicates that $v(t)$ is zero at times $t = \frac{1}{3}$ and $t = 1$. Moreover, it is positive when $t < \frac{1}{3}$ or when $t > 1$ and negative elsewhere. Thus, the particle moves to the right for values of $t < \frac{1}{3}$ and then reverses direction at $t = \frac{1}{3}$, moving to the left; it continues to move to the left and then reverses direction again at $t = 1$; it then continues on, moving to the right.

PROBLEM As a cold front approaches your area, the weather station estimates that the temperature $T$ (in degrees) is a function of time $t$ (in hours) after 10 P.M. of that day according to the equation $T(t)=40-4t+\frac{t^2}{10}$, where $0\le t\le 14$. (a) What will be the temperature at noon the following day, and (b) what is the instantaneous rate of change of the temperature at 3 A.M. and at 10 A.M. of the following day?

SOLUTION (a) Noon of the following day is 14 hours after 10 P.M. of the given day, so $T(14)=40-4(14)+\frac{14^2}{10}=3.6$ degrees.

(b) The instantaneous rate of change (in degrees per hour) in temperature $T$ with respect to $t$, the time after 10 P.M. of the given day, is the derivative, $T'(t)=-4+\frac{t}{5}$. Thus, at 3 A.M., which is 5 hours after 10 P.M., the instantaneous rate of change of the temperature is $T'(5)=-4+\frac{5}{5}=-3$ degrees per hour; at 10 A.M, which is 12 hours after 10 P.M., the instantaneous rate of change of the temperature is $T'(12)=-4+\frac{12}{5}=-1.6$ degrees per hour.

*Solve the following.*

1. A forest fire spreads so that after $t$ hours $f(t)=80t-20t^2$ acres are burning. What is the rate of growth of acreage burning after 1½ hours?
2. The velocity of a thrown ball as a function of time $t$ is given by $v(t) = 80 - 32t$ (feet/second) after being released. What is the acceleration of the ball as a function of time?
3. It is estimated that a shop worker can make $y$ castings $x$ hours after coming to work at 7 A.M. according to the equation $y=3x+8x^2-x^3$. At what rate (castings per hour) is the worker making castings at 9 A.M. of a given day?
4. A pool ball is hit and travels in a straight line. Suppose $s(t)=100t^2+100t$ is the distance (in centimeters) of the ball from its initial position at $t$ seconds. At what velocity is the ball traveling when the ball has traveled 39 centimeters?
5. A particle moves in a horizontal line according to the formula $s(t)=t^4-6t^3+12t^2-10t+3$, where $s$ is the position of the particle at time $t$. Discuss the motion of the particle. (Hint: Factor the derivative.)
6. A particle moves in a straight line according to the formula $s(t)=\frac{t^3}{2}-2t$, where $s$ is the position of the particle at time $t$ (in seconds). Compare the velocity and acceleration of the particle at the end of 2 seconds.
7. A mathematician gardener found that the rate of yield of his garden was $y=60+24x-\frac{12x^2}{5}$ pints of vegetables per $x$ pounds of compost used. What is the rate of change of yield with respect to the amount of compost when he uses 3 pounds of compost?
8. A stone is dropped from the top of a tower and the location from the starting point $s$ (in feet) of the stone at time $t$ (in seconds) is given by the equation $s(t)=-16t^2$, where direction upward is considered positive. If the building is 256 feet tall, find (a) the velocity and (b) the acceleration of the stone after 2 seconds.
9. A potato is projected vertically upward with an initial velocity of 112 feet/second, and it moves as a function of time $t$ (in seconds) according to the formula $s(t)=112t-16t^2$ where $s(t)$ is the distance (in feet) from the starting point. (a) What is the velocity when $t=3$ seconds, and (b) what is the maximum height the potato will reach?

10. Water is being drained from a commercial catfish pond and the volume $V$ (in gallons) of water in the pond after $t$ minutes is given by $V(t) = 250(1600 - 80t + t^2)$. How fast is the water flowing out of the pond at time $t = 5$ minutes?

## Differentiability and continuity

A **differentiable** function is a function that has a derivative. If $f'(c)$ exists, then $f$ is differentiable at $c$; otherwise, $f$ does not have a derivative at $c$.

If a function $f$ is differentiable at $c$, then $f$ is continuous at $c$; in other words, *differentiability implies continuity*. Therefore, if $f$ is *not* continuous at $c$, then $f$ is also *not* differentiable at $c$. *Caution:* Continuity does *not* imply differentiability. A function can be continuous at a point $a$ even though $f'(x)$ does not exist at $a$. This circumstance occurs when there is a cusp (a sharp corner) or a vertical tangent line at $(a, f(a))$. A good example is the continuous function $f(x) = |x|$ for which the derivative does not exist at 0. A graph is shown in Figure 10.1.

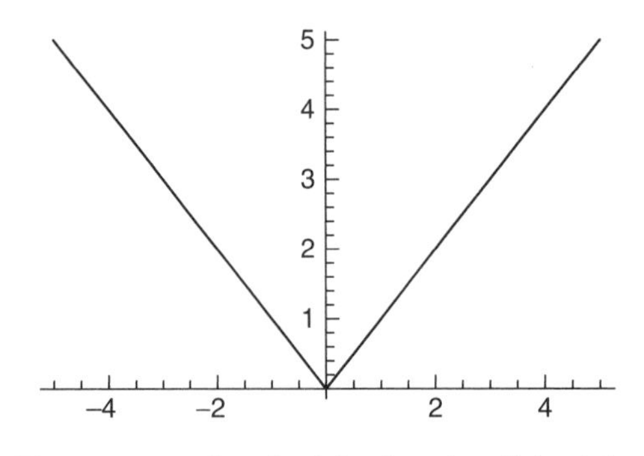


**Figure 10.1** Graph of the function $f(x) = |x|$ for which the derivative does not exist at 0

PROBLEM Show that the function $f(x) = x^{\frac{2}{3}}$ is continuous at $x = 0$, but is not differentiable at $x = 0$.

SOLUTION To investigate we consider $\lim_{x \to 0} f(x) = \lim_{x \to 0} x^{\frac{2}{3}} = 0 = f(0)$, which shows that the function is continuous at $x = 0$. Also, $f'(0) = \lim_{x \to 0} \frac{f(x) - f(0)}{x - 0} = \lim_{x \to 0} \frac{x^{\frac{2}{3}}}{x} = \lim_{x \to 0} \frac{1}{x^{\frac{1}{3}}}$, which does not exist, indicating that $f$ is not differentiable at $x = 0$. Therefore, the function $f(x) = x^{\frac{2}{3}}$ is continuous at $x = 0$, but is not differentiable at $x = 0$.

PROBLEM Determine the values of $x$ for which $f(x) = [x]$, the greatest integer function, is not differentiable.

SOLUTION The function $f(x) = [x]$ has **jump discontinuities** at integer values for $x$; that is, at integer values the left-hand and right-hand limits exist and are finite, but they are different. For instance, if $x$ is less than an integer $n$, as $x$ gets close to $n$ from the left, $f(x) = n - 1$, but if $x$ is greater than $n$, as $x$ gets close to $n$ from the right, $f(x) = n$. Thus the greatest integer function is not differentiable at integer values for $x$. Between non-integer values the function is constant and, thus, differentiable there; in fact, $f'(x) = 0$ at those values.

EXERCISE

**10·3**

*Solve the following.*

1. Determine the values of $x$ for which $f(x)=\dfrac{x^2-5x+6}{x-3}$ is not differentiable.
2. Show that the derivative of $f(x)=|x|$ does not exist at $x=0$, but that the derivative does exist elsewhere.
3. Show that $f(x)=(x-2)^{\frac{1}{3}}$ is continuous at $x=2$, but is not differentiable at 2.
4. Determine whether $f(x)=\begin{cases} 5-6x & x\le 3 \\ -4-x^2 & x>3 \end{cases}$ is differentiable at $x=3$. (Hint: Consider left- and right-hand limits.)
5. Determine whether $f(x)=\begin{cases} x-2 & x<0 \\ x^2 & x\ge 0 \end{cases}$ is differentiable at $x=0$.

# Increasing and decreasing functions, extrema, and critical points

The derivative of a function is a valuable tool in analyzing its graph. For instance, just knowing the algebraic sign of the derivative at a point gives important information. A **sign diagram** for $f'(x)$ is a diagram along the real line showing the signs for $f'(x)$ between critical numbers for *f*. You can use a sign diagram to predict a rough shape of the graph of *f*.

The following definitions are stated for completeness and as a reminder of the concepts.

1. If *f* is continuous on a closed interval [*a*, *b*] and differentiable on the open interval (*a*, *b*), then (i) *f* is **increasing** on [*a*, *b*] if $f'(x)>0$ on (*a*, *b*); (ii) *f* is **decreasing** on [*a*, *b*] if $f'(x)<0$ on (*a*, *b*); and (iii) *f* is **constant** on [*a*, *b*] if $f'(x)=0$ on (*a*, *b*).
2. If *f* is defined on an interval containing *c*, *f*(*c*) is a **minimum** (also, called the **absolute minimum**) of *f* in the interval if $f(c)\le f(x)$ for every number *x* in the interval; similarly, *f*(*c*) is a **maximum** (also, called the **absolute maximum**) of *f* in the interval if $f(c)\ge f(x)$ for every number *x* in the interval. The minimum and maximum values of a function in an interval are the **extreme values**, or **extrema**, of the function in the interval.
3. The number $f(c)$ is a **relative minimum** of a function *f* if there exists an open interval containing *c* in which $f(c)$ is a minimum; similarly, the number $f(c)$ is a **relative maximum** of a function *f* if there exists an open interval containing *c* in which $f(c)$ is a maximum. If $f(c)$ is a relative minimum or maximum of *f*, it is called a **relative extremum** of *f*.
4. If *c* is a number in the domain of *f*, *c* is called a **critical number** of *f* if either $f'(c)=0$ or $f'(c)$ does not exist. The critical numbers determine points at which $f'(x)$ can change signs; that is, these are the only numbers for which the graph of *f* can have turning points, cusps, or discontinuities. If *c* is a critical number for *f*, then $f(c)$ is a critical value of *f* and the point $(c, f(c))$ is a critical point of the graph.
5. If *f* is continuous and has a relative extremum at *c*, then either $f'(c)=0$ or $f'(c)$ does not exist. However, the converse is not necessarily valid. For example, if $f(x)=x^3$, then $f'(x)=3x^2$ and $f'(0)=0$; but $f(0)=0$ is neither a relative maximum nor a relative minimum of the function.

6. **The Extreme Value Theorem** states that if $f$ is continuous on a closed interval $[a, b]$, then $f$ has both a minimum and a maximum value on $[a, b]$.

All of these ideas taken together are tools that can be used to predict the nature and shape of a graph, especially if graphing tools are not applicable or available. Moreover, if graphing is not needed, these ideas can also be extremely valuable in answering maximum and minimum questions.

The function whose graph is depicted in Figure 10.2 has relative and absolute maximums of 1 at $\frac{\pi}{2}$ and $\frac{5\pi}{2}$, relative and absolute minimums of –1 at $\frac{-\pi}{2}$ and $\frac{3\pi}{2}$, is increasing on the intervals $\left[\frac{-\pi}{2}, \frac{\pi}{2}\right]$ and $\left[\frac{3\pi}{2}, \frac{5\pi}{2}\right]$, and is decreasing on the intervals $\left[-\pi, \frac{-\pi}{2}\right], \left[\frac{\pi}{2}, \frac{3\pi}{2}\right]$, and $\left[\frac{5\pi}{2}, 3\pi\right]$. The critical numbers are $-\frac{\pi}{2}, \frac{\pi}{2}, \frac{3\pi}{2}$, and $\frac{5\pi}{2}$ at which the derivative is 0.

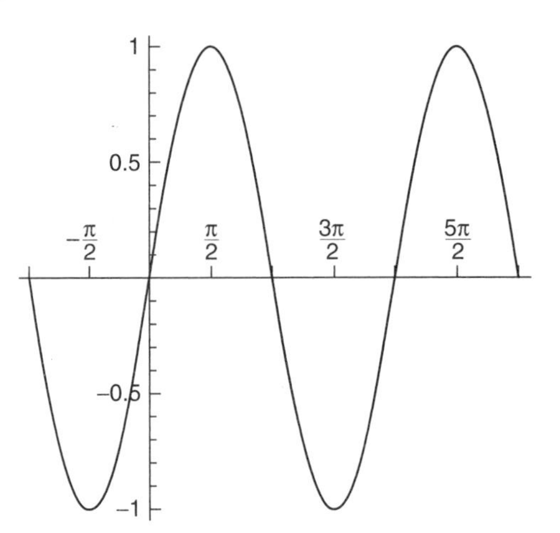


**Figure 10.2** Graph illustrating maxima and minima

A sign diagram for the graph of the function described above would be as illustrated in Figure 10.3.

| $f'(x)$ | – | 0 | + | 0 | – | 0 | + | 0 | – |
|---|---|---|---|---|---|---|---|---|---|
| | | $-\pi/2$ | | $\pi/2$ | | $3\pi/2$ | | $5\pi/2$ | |

**Figure 10.3** Sign graph for the graph of the function in Figure 10.2

The following examples are designed to solidify the concepts above and to give you practice in approaching problems dealing with maximum and minimum ideas.

PROBLEM Given $f(x) = x^3 - 6x^2 + 9x + 1$. (a) Find the critical numbers; (b) find the critical values; and (c) determine where the function is increasing and decreasing.

SOLUTION Differentiating, you have $f'(x) = 3x^2 - 12x + 9$. (a) Setting $f'(x) = 0$, you obtain $3x^2 - 12x + 9 = 3(x-1)(x-3) = 0$. Thus $x = 1$ and $x = 3$ are the critical numbers for $f$. (b) The critical values are $f(1) = 5$ and $f(3) = 1$. (c) When $x < 1, f'(x)$ is positive and so $f$ is increasing for values of $x$ less than 1; when $1 < x < 3, f'(x)$ is negative and so $f$ is decreasing when $x$ is between 1 and 3; when $x > 3, f'(x)$ is positive and so $f$ is increasing for values of $x$ greater than 3.

In the problem above, since $f'(1) = 0 = f'(3)$, the function has horizontal tangents at 1 and 3, and, thus, possibly a relative or absolute maximum or minimum at one or both of these points. This conjecture for the point $x = 1$ can be investigated by evaluating the function at nearby points such as $f(.99) \approx 4.999699$ and $f(1.01) \approx 4.999701$, which seems to indicate a relative maximum at $x = 1$. However, this approach can result in erroneous results because the real numbers are dense and other close-by numbers may give different results.

The following theorems give you analytical tools for making positive decisions regarding maximums and minimums.

The **First Derivative Test** provides that if $c$ is a critical number of a function $f$ that is continuous on an open interval $(a, b)$ containing $c$, then (i) if $f'(x)$ changes sign from negative to positive at $c$, then $f(c)$ is a relative minimum of $f$; and (ii) if $f'(x)$ changes sign from positive to negative at $c$, then $f(c)$ is a relative maximum of $f$.

The **Second Derivative Test** provides that if $f'(c) = 0$ and $f''(c)$ exists on an open interval containing $c$, then (i) $f(c)$ is a relative minimum of $f$ if $f''(c) > 0$; and (ii) $f(c)$ is a relative maximum of $f$ if $f''(c) < 0$. If $f''(c) = 0$, the test is inconclusive.

PROBLEM Given $f(x) = 2x^3 - 9x^2 + 2$, find the critical points and the relative extrema of the function.

SOLUTION Set the first derivative $f'(x) = 6x^2 - 18x = 6x(x - 3) = 0$ to obtain $x = 0$ and $x = 3$ as critical points. Observe that when $x < 0$ or if $x > 3$, $f'(x)$ is positive, and that when $0 < x < 3$, $f'(x)$ is negative. Consequently, by the First Derivative Test, $f(0) = 2$ is a relative maximum because $f'(x)$ changes sign from positive to negative at 0, and $f(3) = -25$ is a relative minimum because $f'(x)$ changes sign from negative to positive at 3.

PROBLEM Given $f(x) = 2x^3 - 9x^2 + 2$, find the critical points and the relative extrema of the function.

SOLUTION Set $f'(x) = 6x^2 - 18x = 6x(x - 3) = 0$ to obtain $x = 0$ and $x = 3$ as critical points. Next, evaluate the second derivative $f''(x) = 12x - 18 = 6(2x - 3)$ at the critical points to obtain $f''(0) = -18$ and $f''(3) = 18$. Thus, by the Second Derivative Test, $f(0) = 2$ is a relative maximum because $f''(0) = -18 < 0$, and $f(3) = -25$ is a relative minimum because $f''(3) = 18 > 0$, which is the same result as was obtained in the first *problem*.

**Note:** The Second Derivative Test is usually invoked if the second derivative is a rather simple calculation. In many cases, it is simpler to use the First Derivative Test than to calculate the second derivative and then test it. Experience with these tests is probably the best way to determine which test to use at any given time.

PROBLEM Given $f(x) = \frac{x^5}{5} - \frac{5x^3}{3} + 4x + 1$. (a) Find the critical numbers and critical values; (b) determine the intervals over which $f$ is increasing and over which $f$ is decreasing; and (c) identify any relative maxima or minima of $f$.

SOLUTION $f'(x) = x^4 - 5x^2 + 4 = (x^2 - 1)(x^2 - 4) = (x - 1)(x + 1)(x - 2)(x + 2)$ and $f''(x) = 4x^3 - 10x = 2x(2x^2 - 5)$. (a) Set $f'(x) = 0$; to obtain the critical numbers $x = \pm 1$ *and* $x = \pm 2$. Thus, the critical values are $f(1) = \frac{53}{15}$, $f(-1) = -\frac{23}{15}$, $f(2) = \frac{31}{15}$, and $f(-2) = -\frac{1}{15}$.

(b) The function is increasing on the intervals $(-\infty,-2],[-1,1]$, and $[2,\infty)$ because the first derivative is positive on these intervals and decreasing on the intervals $[-2,-1]$ and $[1,2]$ because the first derivative is negative on these intervals.

(c) $f''(1)<0, f''(-1)>0, f''(2)>0$, and $f''(-2)<0$, therefore, $f(1)=\frac{53}{15}$ is a relative maximum, $f(-1)=-\frac{23}{15}$ is a relative minimum, $f(2)=\frac{31}{15}$ is a relative minimum, and $f(-2)=-\frac{1}{15}$ is a relative maximum.

PROBLEM 20 feet of wire is to be allocated to form two figures that do not touch: an equilateral triangle and a square. How much wire should be allocated for each figure so that the total area enclosed is a maximum?

SOLUTION There are constraints on the problem that cannot be ignored. These constraints are the following: the amount of wire available and the properties of the geometric figures. Let $x$ denote the length of a side of the square, $s$ denote the length of a side of the equilateral triangle, and $T$ denote the total area enclosed by the two figures. Then, $4x+3s=20,\ 0\le x\le\frac{20}{4}=5,\ 0\le s\le\frac{20}{3}$, the area of the square is $x^2$, and the area of the triangle is $\frac{1}{2}(s)\left(\frac{\sqrt{3}}{2}s\right)=\frac{\sqrt{3}}{4}s^2$. Thus, the total area enclosed by the two figures is $T=x^2+\frac{\sqrt{3}}{4}s^2$. To express $T$ as a function of $x$, solve $4x+3s=20$ for $s$ and substitute the result in the equation for $T$. Thus $T(x)=x^2+\frac{\sqrt{3}}{4}\left(\frac{20-4x}{3}\right)^2$. Now, $T'(x)=2x-\frac{2\sqrt{3}}{9}(20-4x)=\left(\frac{18+8\sqrt{3}}{9}\right)x-\frac{40\sqrt{3}}{9}$, which is 0 when $x=\frac{20\sqrt{3}}{9+4\sqrt{3}}\approx 2.175$. Also, $T''(x)=\frac{18+8\sqrt{3}}{9}>0$, so the critical value $T\left(\frac{20\sqrt{3}}{9+4\sqrt{3}}\right)$ is a relative minimum. By the Extreme Value Theorem, $T$ has an absolute maximum on $0\le x\le\frac{20}{4}=5$. Since the maximum does not occur at the critical number, it must occur at one of the end points of the interval. Since $T(0)=\frac{100\sqrt{3}}{9}\approx 19.245$ and $T(5)=20$, the maximum area is achieved when all the wire is used to form the square.

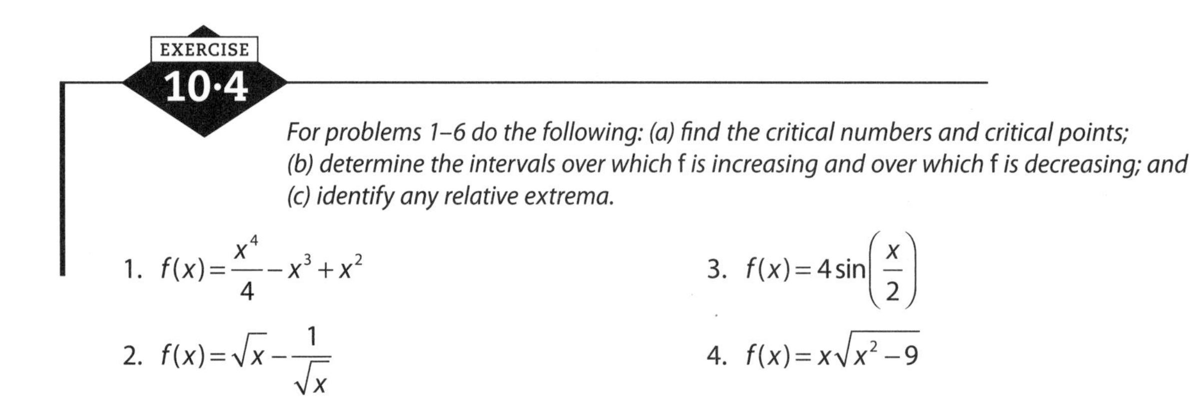

EXERCISE 10·4

*For problems 1–6 do the following: (a) find the critical numbers and critical points; (b) determine the intervals over which f is increasing and over which f is decreasing; and (c) identify any relative extrema.*

1. $f(x)=\frac{x^4}{4}-x^3+x^2$
2. $f(x)=\sqrt{x}-\frac{1}{\sqrt{x}}$
3. $f(x)=4\sin\left(\frac{x}{2}\right)$
4. $f(x)=x\sqrt{x^2-9}$

5. $f(x)=\begin{cases} 2x+9 & \text{if } x\leq -2 \\ x^2+1 & \text{if } x>-2 \end{cases}$

6. $f(x)=2-3(x-4)^{\frac{2}{3}}$

7. Find $a$ and $b$ so that the function $f(x)=x^3+ax^2+b$ will have a relative extreme value at (2, 3).

8. A rancher has 100 feet of wire to make a small pen or pens for chickens. The pens can be in the shape of a square and/or a regular pentagon and do not touch. How much wire should be used for each figure so that the total area enclosed is a maximum? Note: The area of a pentagon is given by the formula $A_p=\dfrac{5s^2\cot(36^\circ)}{2}$ where s is the length of a side.

9. Assume that the amount of money deposited in a bank is proportional to the square of the interest rate the bank pays on this money. Furthermore, the bank can reinvest this money at 9%. Find the interest rate the bank should pay to maximize its profit. (Use the simple interest formula.)

10. A cylindrical glass jar has a flat pewter top. The top costs three times as much as the glass per unit area. Find the proportions, in terms of the height $h$ and radius $r$ of the jar, of the least costly jar that holds a given volume $V$.

# Concavity and points of inflection

If $f$ is a function whose first and second derivatives exist on some open interval containing the number $c$, then (i) the graph of $f$ is **concave upward** at $(c, f(c))$ if $f''(c)>0$, and (ii) the graph of $f$ is **concave downward** at $(c, f(c))$ if $f''(c)<0$.

The point $(c, f(c))$ is a **point of inflection** if the concavity of the graph of $f$ changes at $(c, f(c))$ and if the graph of $f$ has a tangent there.

**Note:** As you work through the examples and exercises in this section, you will find it helpful to know that the graph of a function has a **vertical tangent** at the point $(x_0, f(x_0))$ if and only if $f'(x)$ approaches $\infty$ or $-\infty$ as $x$ approaches $x_0$.

The function whose graph is depicted in Figure 10.4 is concave up on $[-\pi,0]$, concave down on $[0,\pi]$, and concave up on $[\pi,2\pi]$. There are points of inflection at $(0,0)$ and $(\pi,-\pi)$. The function has a horizontal tangent at $(0,0)$.

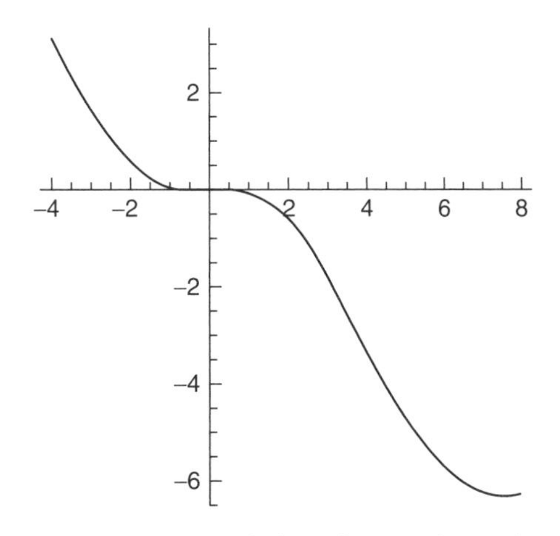


**Figure 10.4** Depiction of concavity and points of inflection

PROBLEM Determine the concavity and points of inflection of the graph of $f(x)=x^4-6x^2-3x-5$.

SOLUTION $f'(x)=4x^3-12x-3$ and $f''(x)=12x^2-12$. Solving $f''(x)=0$ yields $x=\pm 1$. For $x<-1$ or $x>1$, $f''(x)>0$ indicating the graph is concave up when $x<-1$ or $x>1$; for $-1<x<1$, $f''(x)<0$ indicating the graph is concave down.

Consequently, the points $(-1,-7)$ and $(1,-13)$ are points of inflection.

PROBLEM Determine the concavity and points of inflection of the graph of $f(x)=2x^6-3x^4$.

SOLUTION $f'(x)=12x^5-12x^3$ and $f''(x)=60x^4-36x^2=12x^2(5x^2-3)$. Now, the algebraic sign of the second derivative is positive when $x<-\sqrt{\frac{3}{5}}$ or $x>\sqrt{\frac{3}{5}}$ and the sign is negative when $-\sqrt{\frac{3}{5}}<x<\sqrt{\frac{3}{5}}$, so the concavity is upward in the first two cases and downward in the latter case. Hence, the points of inflection occur where the concavity changes, namely, at $\left(-\sqrt{\frac{3}{5}},-\frac{81}{125}\right)$ and $\left(\sqrt{\frac{3}{5}},-\frac{81}{125}\right)$. The graph will be shaped somewhat like a W with zeros at $x=0$ and $x=\pm\sqrt{\frac{3}{2}}$.

PROBLEM Determine the concavity and inflection points of the graph of $f(x)=\begin{Bmatrix} 4-x^2 & \text{if } x\le 1 \\ x^2+2 & \text{if } x>1 \end{Bmatrix}$.

SOLUTION $f'(x)=-2x$ when $x<1$ and $f'(x)=2x$ when $x>1$; but since $\lim\limits_{x\to 1^+}\frac{f(x)-f(1)}{x-1}=\lim\limits_{x\to 1^+}\frac{(x^2+2)-(4-1)}{x-1}=\lim\limits_{x\to 1^+}\frac{x^2-1}{x-1}=\lim\limits_{x\to 1^+}\frac{(x-1)(x+1)}{x-1}=\lim\limits_{x\to 1^+}(x+1)=2$ and $\lim\limits_{x\to 1^-}\frac{f(x)-f(1)}{x-1}=\lim\limits_{x\to 1^-}\frac{(4-x^2)-(4-1)}{x-1}=\lim\limits_{x\to 1^-}\frac{1-x^2}{x-1}=-2$ the derivative does not exist at $x=1$. Moreover, there is no vertical tangent at (1, 3), because the derivative does not approach $\infty$ or $-\infty$ as $x$ approaches 1. Thus, (1, 3) is not a point of inflection.

PROBLEM Determine the concavity and inflection points of the graph of $y=x^{\frac{1}{3}}$.

SOLUTION $y'=\frac{1}{3x^{\frac{2}{3}}}$ and $y''=-\frac{2}{9x^{\frac{5}{3}}}$. The second derivative is positive when $x<0$ and is negative when $x>0$. Since $y'=\frac{1}{3x^{\frac{2}{3}}}$, the derivative approaches infinity as $x$ approaches 0 and, therefore, does not exist at $x=0$. However, the graph has a vertical tangent at (0, 0), and so (0, 0) is a point of inflection.

**Note:** A skillful technique for finding candidates for inflection points is to find the zeros of $f''$ and test its algebraic sign on the left and right of the zeros.

EXERCISE

10·5

*In problems 1–9, determine the concavity and points of inflection of the graph of the function.*

1. $f(x) = x^4 - 6x + 2$
2. $f(x) = x^4 - 6x^3 + 12x^2 - 8x$
3. $g(x) = 3x + 7$
4. $y = 3x + (x+2)^{\frac{3}{5}}$
5. $f(x) = \begin{cases} x^2 + 1 \text{ if } x < 2 \\ 7 - x^2 \text{ if } x \geq 2 \end{cases}$
6. $f(x) = \begin{cases} x^3 \text{ if } x < 0 \\ x^4 \text{ if } x \geq 0 \end{cases}$
7. $f(x) = 2\sin(3x)$ for $x$ in $[-\pi, \pi]$
8. $y = (x-1)^{\frac{1}{3}} - 2$
9. $y = x^4 - 18x^2 + 1$
10. For $h(x) = ax^3 + bx^2 + cx + d$ find values for *a*, *b*, *c*, and *d* so that there is a point of inflection at (1, –1) and a relative maximum at (0, 3)

# Mean Value Theorem

The **Mean Value Theorem (MVT)** states that if the function *f* is continuous in the closed interval $[a, b]$ and if $f'(x)$ exists on the open interval $(a, b)$, then there exists a number *c* in $(a, b)$ such that $f(b) - f(a) = (b-a)f'(c)$.

This theorem is valuable for so many purposes that it is a good exercise to become familiar with some of its nuances. For instance, the theorem guarantees that the number *c* exists, but notice that the theorem does not specify the value of *c*. Many of the problems associated with this concept involve finding values of *c*. On the other hand, in some cases, it may be sufficient just to know that *c* exists. You might also observe that the final equation of the theorem can be rewritten as $f'(c) = \dfrac{f(b) - f(a)}{b - a}$.

A graphical picture of this concept is given in Figure 10.5, where the upper tangent is the line through (1, 3) with slope –2, which is the same as the slope of the secant line that connects the points (0, *f*(0)) and(2, *f*(2)).

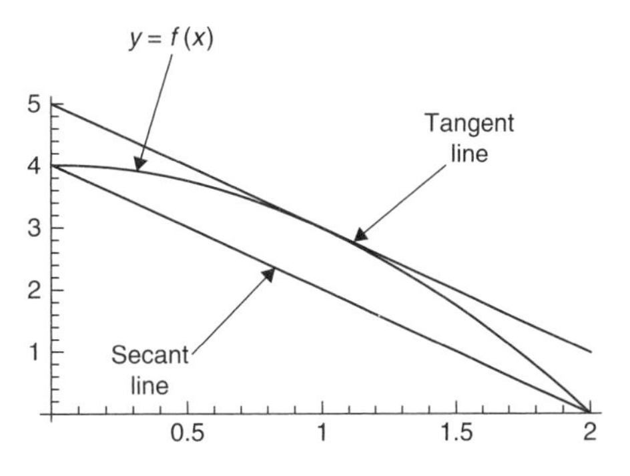


**Figure 10.5** Graphical concept of the Mean Value Theorem

The theorem states that there exists a point between 0 and 2 at which the tangent to the curve has a slope equal to the slope of the depicted secant line.

PROBLEM Find the value of $c$ described by the MVT for $f(x)=x^2$ and the interval [1, 3].

SOLUTION The function $f$ is continuous on [1, 3] and differentiable on (1, 3). Therefore, the MVT applies to the function on the interval [1, 3]. You must find $c$ such that $f'(c)=\frac{f(3)-f(1)}{3-1}=\frac{9-1}{3-1}=4$. Since $f'(x)=2x$, you need to solve $2c=4$ to obtain $c=2$.

PROBLEM Find the value of $c$ described in the MVT for $f(x)=\frac{5}{x+3}$ and the interval $[-2,2]$.

SOLUTION The function $f$ is continuous on $[-2,2]$ and differentiable on $(-2,2)$. Therefore, the MVT applies to the function on the interval $[-2,2]$. Since $f'(x)=\frac{-5}{(x+3)^2}$ and $f(-2)=5$ and $f(2)=1$, you must solve $\frac{-5}{(c+3)^2}=\frac{1-5}{2-(-2)}=-\frac{4}{4}=-1$ or $(c+3)^2=5$, which gives $c=-3\pm\sqrt{5}$. Of these two possible values for $c$, only the value $-3+\sqrt{5}$ lies in the interval $[-2,2]$. Thus, $c=-3+\sqrt{5}$ is the desired value.

PROBLEM Find the value of $c$ described in the MVT for $f(x)=|x|$ and the interval $[-1,3]$.

SOLUTION The function $f$ is continuous on $[-1,3]$; however, when $x<0$, $f'(x)=-1$ and when $x>0$, $f'(x)=1$; so $f'(0)$ does not exist. Therefore, $f$ does not satisfy the hypotheses of the MVT because 0 is in the open interval $(-1,3)$, but $f'(0)$ does not exist. Thus, the MTV does not apply to $f$, so the result is not guaranteed.

PROBLEM Suppose you average 50 mph on at trip that covers 300 miles. Show that at some instant during the trip you traveled at exactly 50 mph.

SOLUTION If $s(t)$ denotes your distance from the starting point at time $t$ then $s'(t)$ is the velocity at time $t$. If you start at time $a$ and end at time $b$ then $s'(c)=\frac{s(b)-s(a)}{b-a}$ for some time $c$ between $a$ and $b$. But $\frac{s(b)-s(a)}{b-a}$ is the average speed, so there is an instant, $c$, such that $s'(c)=50$ mph. (This of course assumes your driving fits the hypotheses of the MVT.)

PROBLEM Show that $\frac{1}{5}<\ln\left(\frac{5}{4}\right)<\frac{1}{4}$.

SOLUTION Let $f(x)=\ln x$ and consider the interval [4, 5]. The function $f$ satisfies the hypotheses of the MVT, so you know there exists $c$ in [4, 5] such that $f'(c)=\frac{1}{c}=\frac{\ln 5-\ln 4}{5-4}=\ln\left(\frac{5}{4}\right)$ or $c=\frac{1}{\ln\left(\frac{5}{4}\right)}$. Substituting this value into $4<c<5$ yields $4<\frac{1}{\ln\left(\frac{5}{4}\right)}<5$, which can be rewritten as the equivalent inequality $\frac{1}{5}<\ln\left(\frac{5}{4}\right)<\frac{1}{4}$, the desired result.

PROBLEM Find the value(s) of $c$ where the mean value is attained for $f(x)=x+\sin x$ on $\left[\frac{\pi}{2}, \pi\right]$.

SOLUTION $f'(x)=1+\cos x$, so you need to solve $1+\cos(c)=\dfrac{(\pi+\sin\pi)-\left(\frac{\pi}{2}+\sin\left(\frac{\pi}{2}\right)\right)}{\pi-\frac{\pi}{2}}=\dfrac{\pi-\left(\frac{\pi}{2}+1\right)}{\frac{\pi}{2}}=1-\dfrac{2}{\pi}$ so that $c=\arccos\left(-\dfrac{2}{\pi}\right)$.

EXERCISE

## 10·6

*Find the value(s) of c where the mean value is attained in problems 1–9.*

1. $f(x)=x^3+x-4$ on $[-1,2]$
2. $f(x)=x+\dfrac{1}{x}$ on $\left[-1,\dfrac{1}{2}\right]$
3. $g(x)=x-\sqrt{x}$ on $[1,4]$
4. $f(x)=\dfrac{x+1}{x-1}$ on $[0,2]$
5. $f(x)=x^3-3x^2$ on $[-1,3]$
6. $h(x)=x^2-x^{\frac{2}{5}}$ on $[-1,1]$
7. $f(x)=x-\cos x$ on $\left[\pi,\dfrac{3\pi}{2}\right]$
8. $f(x)=8x^3+18x^2+3x-7$ on $[-2,1]$
9. $f(x)=\dfrac{\sin x}{1+\cos x}$ on $\left[0,\dfrac{\pi}{2}\right]$
10. Show that $\dfrac{1}{8}<\ln\left(\dfrac{8}{7}\right)<\dfrac{1}{7}$ by considering the function $\ln x$ on $\left[1,\dfrac{8}{7}\right]$

# Applications of the definite integral

## Area of a region under one curve

One of the major advances of calculus was the ability to find areas of plane regions bounded by curves. Euclidean geometry developed formulas and methods for finding areas of plane regions bounded by line segments, but faltered when faced with areas of regions bounded by curves.

If $f$ is a continuous function with $f(x) \geq 0$ on $[a, b]$, then $\int_a^b f(x)dx$ is the area of the region bounded by the curve $y = f(x)$, the $x$-axis, and the lines $x = a$ and $x = b$.

PROBLEM Find the area of the region bounded by the $x$-axis, the lines $x = 4$ and $x = 6$, and the curve $y = x^2 + 2x$.

SOLUTION The function $f$ defined by $y = x^2 + 2x$ is continuous and nonnegative on $[4, 6]$. Thus, the area of the specified region equals $\int_4^6 (x^2 + 2x)\,dx = \left[\frac{x^3}{3} + \frac{2x^2}{2}\right]_4^6 = 70\frac{2}{3}$ square units. This solution assumes, of course, that all measurements are in the same units.

PROBLEM Find the area of the region bounded by $y = x^2 - 2x + 3$, the $x$-axis, and the lines $x = -2$ and $x = 1$.

SOLUTION The function defined by $y = x^2 - 2x + 3$ is continuous and nonnegative on $[-2, 1]$. Thus, the specified area equals $\int_{-2}^{1} (x^2 - 2x + 3)\,dx = \left[\frac{x^3}{3} - \frac{2x^2}{2} + 3x\right]_{-2}^{1} = 15$ square units.

PROBLEM Find the area of the region bounded by the curve $y = \tan^2 x$, the $x$-axis, and the lines $x = 0$ and $x = \frac{\pi}{4}$.

SOLUTION The function defined by $y = \tan^2 x$ is continuous and nonnegative on $\left[0, \frac{\pi}{4}\right]$. Thus, the specified area equals $\int_0^{\frac{\pi}{4}} \tan^2 x\,dx = [\tan x - x]_0^{\frac{\pi}{4}} = \left(1 - \frac{\pi}{4}\right)$ square units.

PROBLEM Find the area of the region bounded by $y=x^2$, the $x$-axis, and the lines $x=0$ and $x=1$.

SOLUTION The function defined by $y=x^2$ is continuous and nonnegative on [0, 1] so the area is given by $\int_0^1 x^2 dx = \left[\frac{x^3}{3}\right]_0^1 = \frac{1}{3}$ square units.

EXERCISE

**11·1**

*Find the area of the region bounded by the indicated curves.*

1. $y=2x^2+2x-24$; $x$-axis; $x=3$; $x=6$
2. $y=\sin x$; $x$-axis; $x=\frac{\pi}{3}$; $x=\frac{2\pi}{3}$
3. $y=8x-2x^2$; $x$-axis; $x=1$; $x=3$
4. $y=\sec^2 x$; $x$-axis; $y$-axis; $x=\frac{\pi}{4}$
5. $y=\sqrt{4x+4}$; $x$-axis; $y$-axis; $x=8$
6. $y=\cos x$; $x$-axis; $y$-axis; $x=\frac{\pi}{6}$

# Area of a region between two curves

If $f$ and $g$ are continuous functions with $f(x)\geq g(x)$ on $[a, b]$, then the area between the two curves is given by $\int_a^b [f(x)-g(x)]\,dx$.

As you can see, the problem of finding areas between curves involves essentially exploiting ideas developed in the first section of this chapter.

PROBLEM Find the area enclosed by the curves $y=f(x)=x^3+x$ and $y=h(x)=\sin x$, the $x$-axis, and the lines $x=\frac{\pi}{2}$ and $x=\pi$.

SOLUTION Both $f$ and $h$ are continuous and nonnegative in $\left[\frac{\pi}{2}, \pi\right]$ and $f(x)\geq h(x)$ on $\left[\frac{\pi}{2}, \pi\right]$. Thus, the specified area is given by $\int_{\pi/2}^{\pi} [(x^3+x)-\sin x]\,dx =$

$$\left[\frac{x^4}{4}+\frac{x^2}{2}+\cos x\right]_{\pi/2}^{\pi} = \left[\frac{\pi^4}{4}+\frac{\pi^2}{2}-1\right]-\left[\frac{\pi^4}{2^6}+\frac{\pi^2}{2^3}\right]=\left(\frac{15\pi^4}{64}+\frac{3\pi^2}{8}-1\right)$$ square units.

PROBLEM Find the area enclosed by the lines $x=0$, $x=1$, the $x$-axis, and the curves $y=f(x)=-x+1$ and $y=g(x)=x^2$.

SOLUTION Solve for the intersection of the two functions by equating the expressions to get $x^2=-x+1$ or $x^2+x-1=0$. The solutions to this quadratic equation are $x=\frac{-1\pm\sqrt{5}}{2}$ and the value in the interval [0, 1] is $x=\frac{\sqrt{5}-1}{2}$. Also, $f$

dominates when $x$ is less than this value and $g$ dominates when $x$ is greater than this value so the specified area equals $\int_0^{\frac{\sqrt{5}-1}{2}} [(-x+1)-x^2]\,dx$ +

$$\int_{\frac{\sqrt{5}-1}{2}}^{1} [x^2-(-x+1)]\,dx = \left[\frac{-x^2}{2}+x-\frac{x^3}{3}\right]_0^{\frac{\sqrt{5}-1}{2}} + \left[\frac{x^3}{3}+\frac{x^2}{2}-x\right]_{\frac{\sqrt{5}-1}{2}}^{1}$$

$$= \left(-\frac{\left(\frac{\sqrt{5}-1}{2}\right)^2}{2}+\left(\frac{\sqrt{5}-1}{2}\right)-\frac{\left(\frac{\sqrt{5}-1}{2}\right)^3}{3}\right)+\left(\frac{1}{3}+\frac{1}{2}-1\right)$$

$$-\left(\frac{\left(\frac{\sqrt{5}-1}{2}\right)^3}{3}+\frac{\left(\frac{\sqrt{5}-1}{2}\right)^2}{2}-\left(\frac{\sqrt{5}-1}{2}\right)\right)=2\left(-\frac{\left(\frac{\sqrt{5}-1}{2}\right)^2}{2}+\left(\frac{\sqrt{5}-1}{2}\right)-\frac{\left(\frac{\sqrt{5}-1}{2}\right)^3}{3}\right)$$

$$-\frac{1}{6}=2\left(-\frac{(3-\sqrt{5})}{4}+\frac{\sqrt{5}-1}{2}-\frac{(\sqrt{5}-2)}{3}\right)-\frac{1}{6}=2\left(-\frac{3}{4}-\frac{1}{2}+\frac{2}{3}+\frac{\sqrt{5}}{4}+\frac{\sqrt{5}}{2}-\frac{\sqrt{5}}{3}\right)-\frac{1}{6}$$

$$=2\left(\frac{5\sqrt{5}-7}{12}\right)-\frac{1}{6}=\frac{5\sqrt{5}-8}{6}\approx 0.530057 \text{ square units.}$$

PROBLEM Find the area of the region between $f(x)=x^2+2$ and $g(x)=1-x$ between $x=0$ and $x=1$.

SOLUTION The functions $f$ and $g$ are continuous on [0, 1]. Moreover, since $x\geq 0,\ x^2+x+1>0$. Thus, $x^2+2>1-x$ and so $f(x)\geq g(x)$. The area of the specified region is

$$\int_0^1[(x^2+2)-(1-x)]\,dx=\int_0^1[x^2+x+1]\,dx=\left[\frac{x^3}{3}+\frac{x^2}{2}+x\right]_0^1=\left(\frac{1}{3}+\frac{1}{2}+1\right)=\frac{11}{6}$$

square units.

PROBLEM Find the area between the curves $y=x-1$ and $y=2x^3-1$ between $x=1$ and $x=2$.

SOLUTION The functions defined by $y=g(x)=x-1$ and $y=f(x)=2x^3-1$ are continuous functions. Furthermore, since $1\leq x\leq 2$ it follows that $x-1\leq 1$ and $2x^3-1\geq 1$ and so $f(x)\geq g(x)$. Thus, the specified area equals $\int_1^2[(2x^3-1)-(x-1)]\,dx=$

$$\int_1^2(2x^3-x)\,dx=\left[\frac{2x^4}{4}-\frac{x^2}{2}\right]_1^2=(8-2)-0=6 \text{ square units.}$$

PROBLEM Find the area of the region bounded by the graphs of $f(x)=x^3-3x+2$ and $g(x)=x+2$.

SOLUTION Consider $f(x)-g(x)=(x^3-3x+2)-(x+2)=x^3-4x=x(x-2)(x+2)$. When $-2\leq x\leq 0$, you have $f(x)-g(x)\geq 0$ or, equivalently, $f(x)\geq g(x)$; and when $0\leq x\leq 2$, you have $f(x)-g(x)\leq 0$ or, equivalently, $f(x)\leq g(x)$. Based on this information, you know the required area is enclosed in two regions. Thus, the area is given by the sum of the two integrals that follow.

$$Area = \int_{-2}^{0} [(x^3 - 3x + 2) - (x + 2)]\,dx + \int_{0}^{2} [(x + 2) - (x^3 - 3x + 2)]\,dx$$

$$= \int_{-2}^{0} (x^3 - 4x)\,dx + \int_{0}^{2} (4x - x^3)\,dx = \left[\frac{x^4}{4} - \frac{4x^2}{2}\right]_{-2}^{0} + \left[\frac{4x^2}{2} - \frac{x^4}{4}\right]_{0}^{2} = 4 + 4 = 8$$

square units.

EXERCISE

**11·2**

*Find the area of the region bounded by the given curves.*

1. $f(x) = 4 - x^2$ and the $x$-axis
2. $y = x^2$ and $y = x + 2$
3. $y = x^2$ and $y = \sqrt{x}$
4. $y = (x + 1)^3$ and $y = x + 1$
5. $y = x^3 + x$; $y = 0$; $x = -1$; $x = 1$
6. $y = 2x + 3$; $y = -x + 6$; $x = 0$; $x = 1$
7. $y = e^x$; $y = e$; $x = 0$
8. $y = x^4 - 2x^3 - x^2 + 2x + 1$; $y = 1$; $x = -1$; $x = 0$
9. $y = 3 - x^2$; $y = 1 - x$; $x = -1$; $x = 1$
10. $y = x^2$; $y = 1$

# Length of an arc

If a function $f$ has a continuous derivative on $[a, b]$, then the length of the arc of the curve $y = f(x)$ between the point $(a, f(a))$ and the point $(b, f(b))$ is given by the formula

$$\text{arc length} = L = \int_{a}^{b} \sqrt{1 + [f'(x)]^2}\,dx$$

On the other hand, if $x = h(y)$ is expressed as a function of y and $h'$ is continuous on the interval $[c, d]$, then $L = \int_{c}^{d} \sqrt{1 + [h'(y)]^2}\,dy$.

PROBLEM Find the length of the arc of the curve $y = f(x) = x^{\frac{2}{3}}$ from the point (1, 1) to the point (8, 4).

SOLUTION The length of the specified arc $= L = \int_{a}^{b} \sqrt{1 + [f'(x)]^2}\,dx = \int_{1}^{8} \sqrt{1 + \left(\frac{2}{3x^{\frac{1}{3}}}\right)^2}\,dx =$

$\int_{1}^{8} \sqrt{1 + \frac{4}{9x^{\frac{2}{3}}}}\,dx = \int_{1}^{8} \sqrt{\frac{9x^{\frac{2}{3}} + 4}{9x^{\frac{2}{3}}}}\,dx = \frac{1}{3}\int_{1}^{8} \frac{\sqrt{9x^{\frac{2}{3}} + 4}}{x^{\frac{1}{3}}}\,dx$. Now if you make the substitution

$u = 9x^{\frac{2}{3}} + 4$ then the integral transforms to $L = \frac{1}{18}\int_{13}^{40} u^{\frac{1}{2}}\,du = \frac{1}{18}\left[\frac{2u^{\frac{3}{2}}}{3}\right]_{13}^{40} =$

$\frac{1}{27}\left(40^{\frac{3}{2}} - 13^{\frac{3}{2}}\right) \approx 7.6$. A simpler integration could be achieved by first solving for $x$ in terms of $y$ and using the appropriate formula.

PROBLEM Find the length of the arc of $f(x) = \frac{2}{3}(1+x^2)^{\frac{3}{2}}$ between $x = 0$ and $x = 3$.

SOLUTION $L = \int_a^b \sqrt{1+[f'(x)]^2}\,dx = \int_0^3 \sqrt{1+\left[(1+x^2)^{\frac{1}{2}}(2x)\right]^2}\,dx = \int_0^3 \sqrt{1+4x^2+4x^4}\,dx =$

$\int_0^3 \sqrt{(1+2x^2)^2}\,dx = \int_0^3 (1+2x^2)\,dx = \left[x + \frac{2x^3}{3}\right]_0^3 = 21.$

PROBLEM Find the length of the arc of $x^2 = 1 - e^y$ between $x = 0$ and $x = \frac{1}{2}$.

SOLUTION First, solve for $y$ in terms of $x$ to get $y = \ln(1-x^2)$. Then apply the formula

to get $L = \int_a^b \sqrt{1+[f'(x)]^2}\,dx = \int_0^{\frac{1}{2}} \sqrt{1+\left(\frac{-2x}{1-x^2}\right)^2}\,dx = \int_0^{\frac{1}{2}} \sqrt{\frac{x^4+2x^2+1}{(1-x^2)^2}}\,dx =$

$\int_0^{\frac{1}{2}} \sqrt{\frac{(x^2+1)^2}{(1-x^2)^2}}\,dx = \int_0^{\frac{1}{2}} \frac{x^2+1}{1-x^2}\,dx = \int_0^{\frac{1}{2}} \left(\frac{2}{1-x^2} - 1\right)dx = \left[\ln\left|\frac{x+1}{x-1}\right| - x\right]_0^{\frac{1}{2}} = \ln 3 - \frac{1}{2}.$

## EXERCISE 11·3

*Find the arc length of the indicated curve on the given interval.*

1. $y = \frac{x^2}{2}$ between $x = -\sqrt{3}$ and $x = 0$
2. $y = 4 - \frac{4x}{9}$ between its $x$ and $y$ intercepts
3. $y = \frac{(x^2+2)^{\frac{3}{2}}}{3}$ on [0, 3]
4. $6xy = y^4 + 3$ from $y = 1$ to $y = 2$
5. $y = \frac{x^4}{4} + \frac{1}{8x^2}$ on [1, 2]
6. $y = \frac{\sqrt{x}(3x-1)}{3}$ on [1, 4]
7. $y = \ln x$ on $[1, \sqrt{3}]$
8. $y = \frac{x^3}{3} + \frac{1}{4x}$ on [1, 3]
9. $y^2 = \frac{x(x-3)^2}{9}$; the length desired is in the first quadrant on [1, 3]
10. $y = 2(x-1)^{\frac{3}{2}}$ on $\left[1, \frac{17}{9}\right]$

## APPENDIX A

# Basic functions and their graphs

A **function** $f$ is a set of ordered pairs $(x, y)$ for which each first element, $x$, is paired with *one and only one* second element, $y$; that is, if $(x, y_1) \in f$ and $(x, y_2) \in f$, then $y_1 = y_2$. The symbol $f(x)$ (read "$f$ of $x$") is commonly used to denote the value (or **image**) of $f$ at $x$. Thus, $(x, y) \in f$ can be expressed as $(x, f(x)) \in f$ or, simply, $y = f(x)$. The set consisting of all the first elements in the ordered pairs contained in $f$ is the **domain** of $f$, and the set of all second elements is the **range** of $f$.

**Linear functions** are defined by equations of the form $y = mx + b$. The domain and range for a linear function are both $\boldsymbol{R}$, the set of real numbers. The only zero is $x = \dfrac{-b}{m}$; that is, the graph crosses the $x$-axis at the point $\left(\dfrac{-b}{m}, 0\right)$, provided $m$ is not zero. The graph is a straight line that has slope $m$ and $y$-intercept $b$. Figure A.1 shows a graph of the linear function $y = 2x + 5$.

The **identity function** is the linear function defined by the equation $y = x$. The identity function maps each $x$-value to an identical $y$-value. The only zero is $x = 0$. The graph passes through the origin, so both the $x$- and $y$-intercepts are zero. Figure A.2 shows the identity function.

**Constant functions** are linear functions defined by equations of the form $y = b$, where $b \in R$. The domain is $R$, and the range is the set $\{b\}$ containing the single element $b$. Constant functions can have either no zeros or infinitely many zeros: If $b \neq 0$, it has no zeros; if $b = 0$, every real number $x$ is a zero. The graph of a constant function is a horizontal line that is $|b|$ units above or below the $x$-axis when $b \neq 0$ and is coincident with the $x$ axis when $b = 0$. Figure A.3 shows the constant function $y = -2$.

**Proportional functions** are linear functions defined by equations of the form $y = kx$, where $k \in R$ is a constant called the **constant of proportionality**. The domain and range are both $R$. The only zero is $x = 0$. The graph passes through the origin, so both the $x$- and $y$-intercept is zero. Figure A.4 shows the proportional function $y = 4x$.

**Quadratic functions** are defined by equations of the form $y = ax^2 + bx + c$, $(a \neq 0)$. The domain is $R$ and the range is a subset of $R$. The zeros are the roots of the quadratic equation $ax^2 + bx + c = 0$. The quantity $b^2 - 4ac$ is called the **discriminant** of the quadratic equation. It determines three cases for the zeros of the quadratic function: If $b^2 - 4ac > 0$, the quadratic function has two real *unequal* zeros; if $b^2 - 4ac = 0$, the quadratic function one real zero (double root); and if $b^2 - 4ac < 0$, the quadratic function has no real zeros.

The graphs of quadratic functions defined by equations of the form $y = ax^2 + bx + c$ are **parabolas**. When $a > 0$, the parabola opens upward and has a **minimum value** at its vertex. When $a < 0$, the parabola opens downward and has

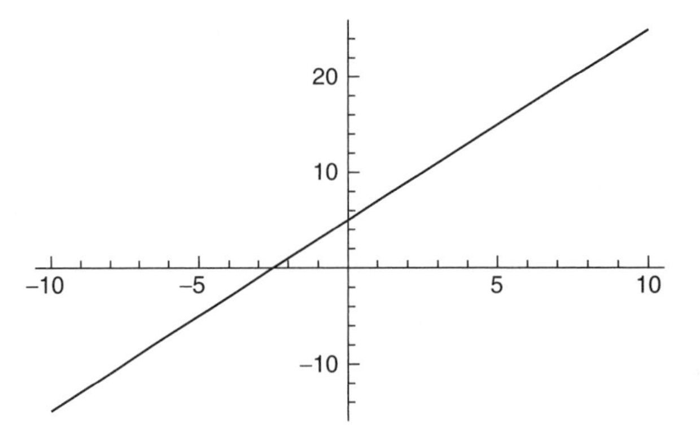


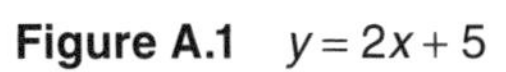

**Figure A.1** $y = 2x + 5$

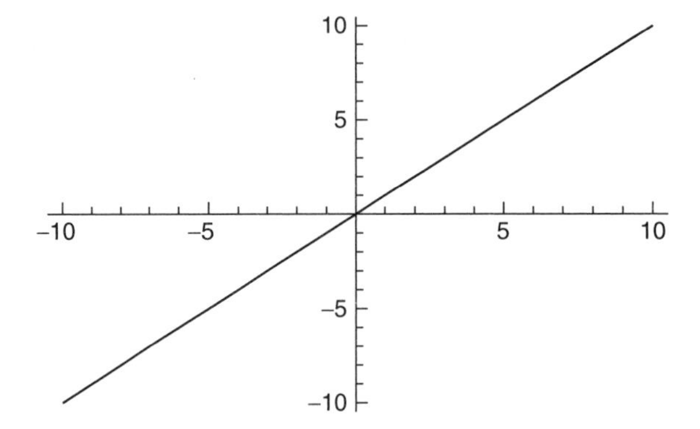

**Figure A.2** $y = x$

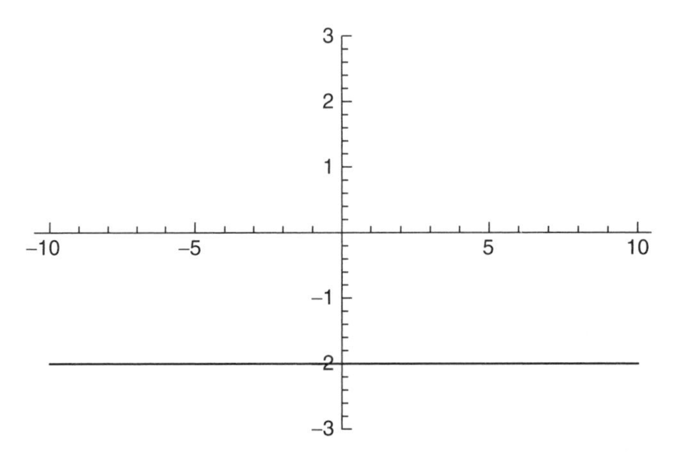


**Figure A.3** $y = -2$

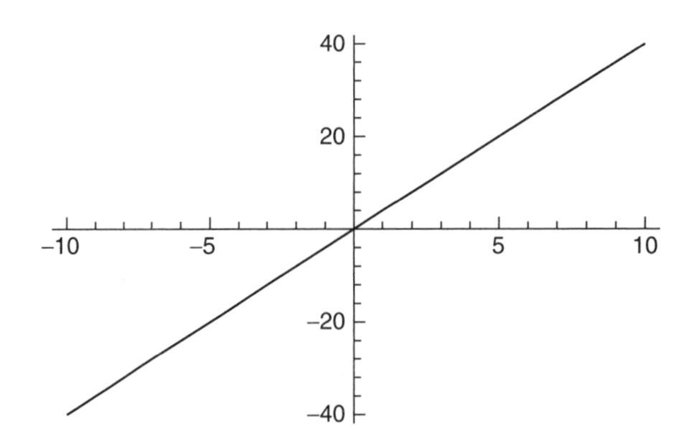


**Figure A.4** $y = 4x$

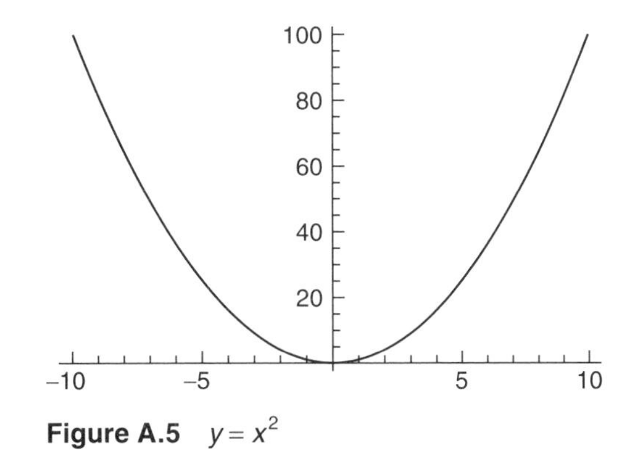


**Figure A.5** $y = x^2$

a **maximum value** at its vertex. The parabola is symmetric about its **axis of symmetry**, a vertical line through its vertex that is parallel to the $y$-axis. Depending on the solution set of $ax^2 + bx + c = 0$, the parabola might or might not intersect the $x$-axis. Three cases occur: If $ax^2 + bx + c = 0$ has *two* real *unequal* roots, the parabola will intersect the $x$-axis at those *two* points; if $ax^2 + bx + c = 0$ has exactly *one* real root, the parabola will be tangent to the $x$-axis at only that *one* point; and if $ax^2 + bx + c = 0$ has no real roots, the parabola will *not* intersect the $x$-axis. Figure A.5 shows the quadratic function $y = x^2$.

**Polynomial functions** are defined by equations of the form $y = P(x) = a_n x^n + a_{n-1} x^{n-1} + \cdots + a_1 x + a_0$, where $n$ is a nonnegative integer and $a_n \neq 0$. The degree of the polynomial is $n$. Linear and quadratic functions are polynomial functions of degree one and two, respectively. The domain for any polynomial function is $R$. When $n$ is odd, the range is $R$. When $n$ is even, the range is a subset of $R$. The zeros, if any, are the solutions of the equation $P(x) = 0$. A number $r$ is a zero of $y = P(x)$ if it is a root of the equation $P(x) = 0$. If $P(r) = 0$ and $r \in R$, the graph of $y = P(x)$ intersects the $x$-axis at $r$. The graph has $y$-intercept $P(0)$. Figure A.6 shows the polynomial function $y = x^3$ (called the **cubic function**) and Figure A.7 shows the polynomial function $y = x^4 - 2x^2 - x + 1$.

Some useful theorems to know about polynomial functions are the following:

1. If $a, b \in R$ such that $P(a)$ and $P(b)$ have opposite signs, then $P$ has at least one zero between $a$ and $b$.
2. **Fundamental Theorem of Algebra.** Over the complex numbers, every polynomial of degree $n \geq 1$ has at least one zero. It follows that if you allow complex roots and count a root again each time it occurs more than once, a polynomial of degree $n$ has exactly $n$ roots.
3. **Factor Theorem.** $P(r) = 0$ if and only if $x - r$ is a factor of $P(x) = a_n x^n + a_{n-1} x^{n-1} + \cdots + a_1 x + a_0$.

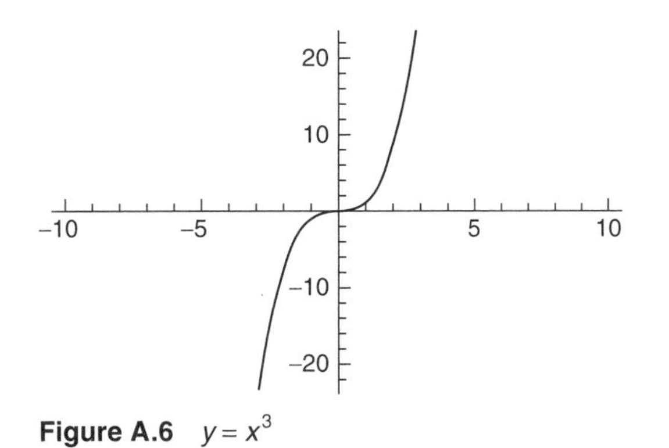


**Figure A.6** $y = x^3$

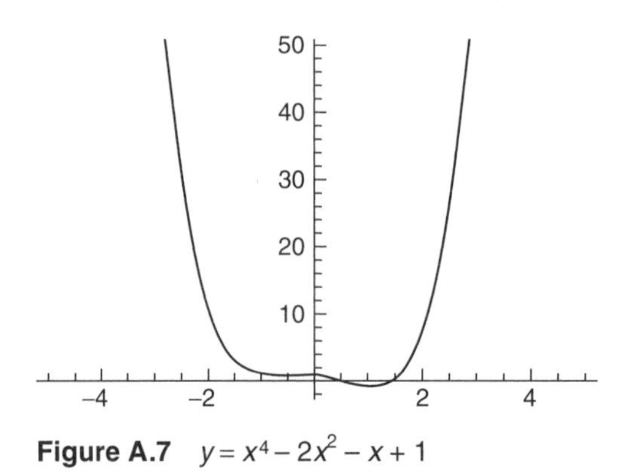

**Figure A.7** $y = x^4 - 2x^2 - x + 1$

4. **Remainder Theorem.** If a polynomial $P(x) = a_n x^n + a_{n-1}x^{n-1} + \cdots + a_1 x + a_0$ is divided by $x - a$, the remainder is $P(a)$.
5. **Descartes' Rule of Signs.** If $y = P(x)$ is a polynomial with real coefficients, then the number of positive real roots of $P(x) = 0$ is either the number of sign changes, from left to right, occurring in the coefficients of $P(x)$, or else is less than this number by an even number. Similarly, the number of negative real roots of $P(x) = 0$ is either the number of sign changes, from left to right, occurring in the coefficients of $P(-x)$, or else is less than this number by an even number.

**Rational functions** are defined by equations of the form $y = \frac{P(x)}{Q(x)} = \frac{a_n x^n + a_{n-1}x^{n-1} + \cdots + a_1 x + a_0}{b_m x^m + b_{m-1}x^{m-1} + \cdots + b_1 x + b_0}$, where $P(x)$ and $Q(x)$ are polynomials and $Q(x) \neq 0$. The domain is $\{x \in R \mid Q(x) \neq 0\}$. The range is a subset of $R$. The zeros, if any, occur at $x$ values for which $y = 0$. Since $y = \frac{P(x)}{Q(x)}$ is not defined when $Q(x) = 0$, graphing the function usually proceeds by determining asymptotes of the function. The **vertical asymptotes**, if any, will be located at values for $x$ (if any) for which $Q(x) = 0$. Use the following guidelines to identify **horizontal asymptotes**: if $n < m$, then the $x$-axis ($y = 0$) is a horizontal asymptote; if $n = m$, then $y = \frac{a_n}{b_n}$, is a horizontal asymptote; and if $n > m$, $y = \frac{P(x)}{Q(x)}$ will *not* have a horizontal asymptote. However, if $n = m + 1$, then $y = \frac{P(x)}{Q(x)}$ will have an **oblique asymptote**, which you can find by dividing $P(x)$ by $Q(x)$. Figure A.8 shows the rational function $y = \frac{x+1}{x-1}$.

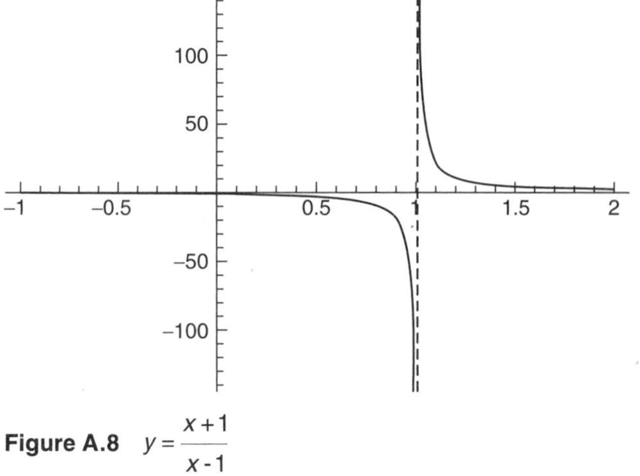

**Figure A.8** $y = \frac{x+1}{x-1}$

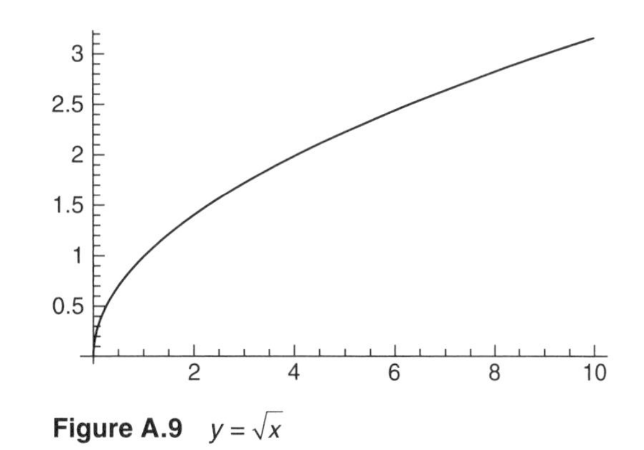

**Figure A.9** $y = \sqrt{x}$

**Square root functions** are defined by equations of the form $y = \sqrt{ax+b}$. The domain is $\{x \in R \mid ax+b \geq 0\}$. The range is $\{y \in R \mid y \geq 0\}$. The graph is nonnegative with the only zero at $x = -\frac{b}{a}$. Figure A.9 shows the square root function $y = \sqrt{x}$.

**Absolute value functions** are defined by equations of the form $y = |ax+b|$. The domain is $R$, and the range is $\{y \in R \mid y \geq 0\}$. The only zero occurs at $x = -\frac{b}{a}$, and the $y$-intercept is located at $|b|$. Technically, $y = |ax+b|$ is a **piecewise function** because you can write it as $y = \begin{cases} ax+b \text{ if } x \geq -\frac{b}{a} \\ -(ax+b) \text{ if } x < -\frac{b}{a} \end{cases}$. Figure A.10 shows the absolute value function $y = |x|$.

The **greatest integer function** is defined by $y = [x]$, where the brackets denote to find the greatest integer $n$ such that $n \leq x$. The domain is $R$, and the range is $\mathbb{Z}$, the set of integers. The zeros consists of all the numbers in the interval [0, 1). Figure A.11 shows the greatest integer function $y = [x]$.

**Exponential functions** are defined by equations of the form $y = b^x (b \neq 1, b > 0)$, where $b$ is the **base** of the exponential function. The domain is $R$, and the range is $\{y \in R \mid y > 0\}$. The graph of $y = b^x$ does not cross the $x$-axis, so there are no zeros. The graph passes through (0, 1) and (1, $b$). The $x$-axis is a horizontal asymptote. The function is increasing if $b > 1$ and decreasing if $0 < b < 1$.

Two important exponential functions are defined by $y = 10^x$, with base 10, and $y = e^x$, the **natural exponential function**, with base $e$, the irrational number whose rational decimal approximation is 2.718281828 (to nine digits). Figure A.12 shows the exponential function $y = e^x$.

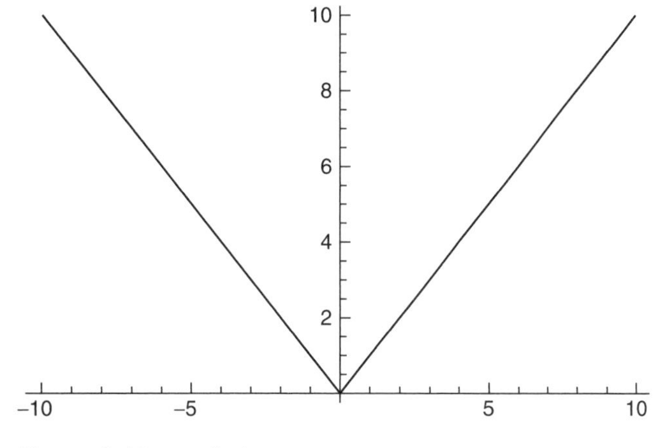

**Figure A.10** $y = |x|$

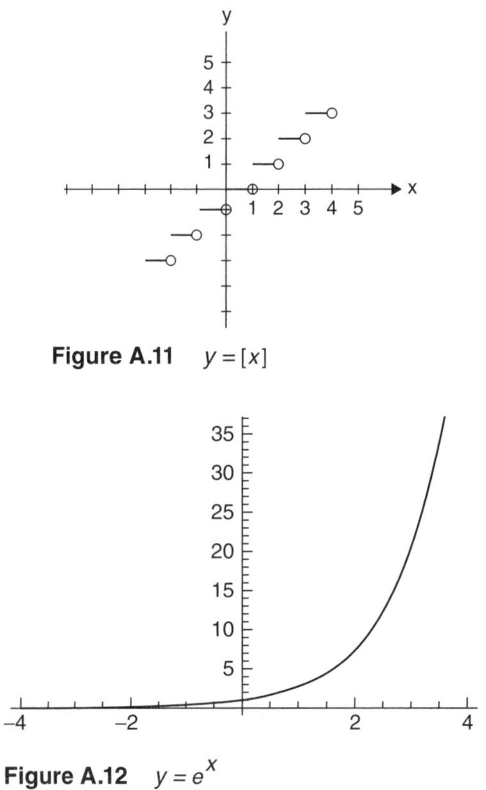


**Figure A.11** $y = [x]$

**Figure A.12** $y = e^x$

**Logarithmic functions** are defined by equations of the form $y = \log_b x$, where $b$ is the **base** of the logarithmic function, $(b \neq 1, b > 0)$ such that $b^y = x\,(x > 0)$. The domain is $\{x \in R \mid x > 0\}$, and the range is $R$. The only zero occurs at $x = 1$. The graph passes through (1, 0) and ($b$, 1). The $y$-axis is a vertical asymptote. The function is increasing if $b > 1$ and is decreasing if $0 < b < 1$. Figure A.13 shows the logarithm function $y = \ln x$.

For a given base, the logarithmic function is the **inverse** of the corresponding exponential function, and conversely. The logarithm function $y = \log_{10} x$ (**common logarithmic function**) is the inverse of the exponential function $y = 10^x$. The logarithm function $y = \ln x$ (**natural logarithmic function**) is the inverse of the exponential function $y = e^x$.

The standard **sine function** is defined by $y = \sin x$. The domain is $R$, and the range is $\{y \in R \mid -1 \leq y \leq 1\}$. The zeros occur at $x = k\pi$, where $k \in \mathbb{Z}$. The graph is periodic with period $= 2\pi$. The curve is a sinusoidal wave that has amplitude $= 1$. The general sine function is defined by $y = a \sin b(x - h) + k$, which is "centered" at ($h$, $k$), has amplitude $= |a|$, and period $= \dfrac{2\pi}{|b|}$. Figure A.14 shows the sine function $y = \sin x$.

The standard **cosine function** is defined by $y = \cos x$. The domain is $R$, and the range is $\{y \in R \mid -1 \leq y \leq 1\}$. The zeros occur at $x = \dfrac{\pi}{2} + k\pi$, where $k \in \mathbb{Z}$. The graph is periodic with period $= 2\pi$. The curve is a sinusoidal wave that has amplitude $= 1$. The general cosine function is defined by $y = a \cos b(x - h) + k$, which is "centered" at ($h$, $k$), has amplitude $= |a|$, and period $= \dfrac{2\pi}{|b|}$. Figure A.15 shows the cosine function $y = \cos x$.

The standard **tangent function** is defined by $y = \tan x$. The domain is $\left\{x \in R \,\middle|\, x \neq \dfrac{\pi}{2} + k\pi\right\}$, and the range is $R$. The zeros occur at $x = k\pi$, where $k \in \mathbb{Z}$. The graph is periodic with period $= \pi$. The curve has vertical asymptotes at $x = \dfrac{\pi}{2} + k\pi$. The general tangent function is defined by $y = a \tan b(x - h) + k$, which is "centered" at ($h$, $k$) and has period $= \dfrac{\pi}{|b|}$. Figure A.16 shows the tangent function $y = \tan x$.

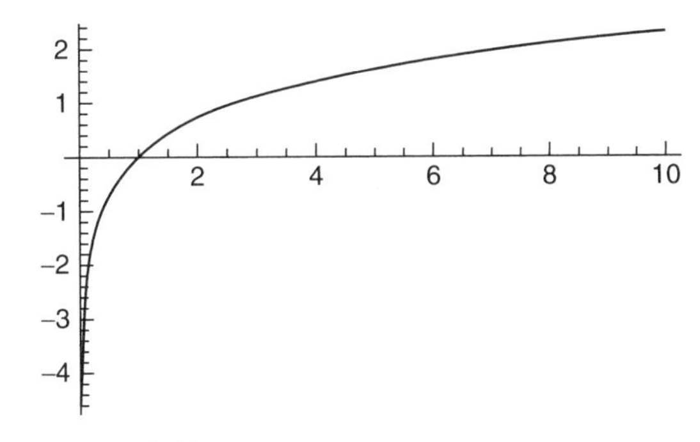


**Figure A.13** $y = \ln x$

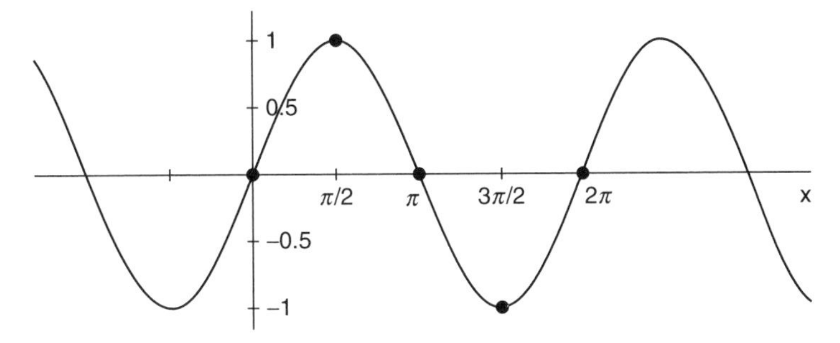


**Figure A.14** $y = \sin x$

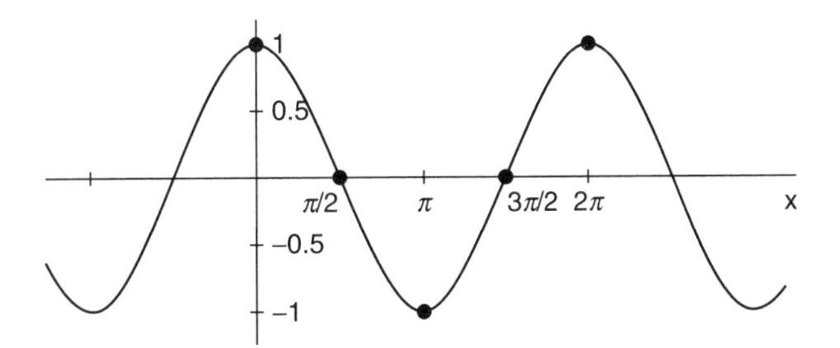


**Figure A.15** $y = \cos x$

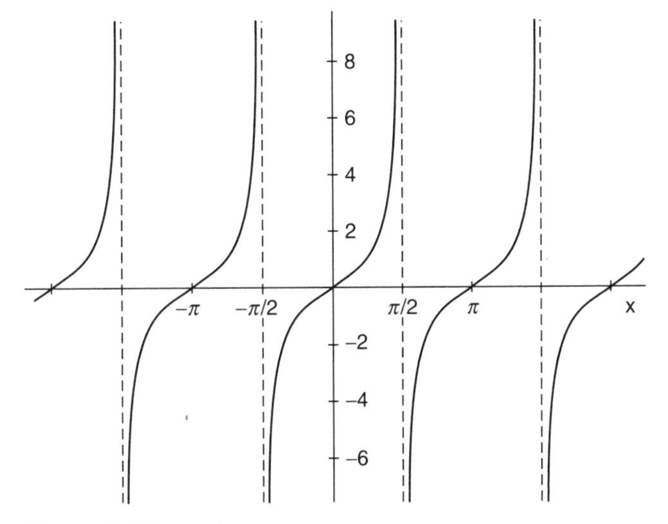


**Figure A.16** $y = \tan x$

## APPENDIX B

# Basic differentiation formulas and rules

1. $\frac{d}{dx}(c)=0$
2. $\frac{d}{dx}(x)=1$
3. $\frac{d}{dx}(mx+b)=m$
4. $\frac{d}{dx}(mu+b)=m\cdot\frac{du}{dx}$
5. $\frac{d}{dx}(x^n)=nx^{n-1}$
6. $\frac{d}{dx}(u^n)=nu^{n-1}\cdot\frac{du}{dx}$
7. $\frac{d}{dx}(e^x)=e^x$
8. $\frac{d}{dx}(e^u)=e^u\cdot\frac{du}{dx}$
9. $\frac{d}{dx}(\ln x)=\frac{1}{x}$
10. $\frac{d}{dx}(\ln kx)=\frac{1}{x}$
11. $\frac{d}{dx}(\ln u)=\frac{1}{u}\cdot\frac{du}{dx}$
12. $\frac{d}{dx}(b^x)=(\ln b)b^x,\ b\neq 1$
13. $\frac{d}{dx}(b^u)=(\ln b)b^u\cdot\frac{du}{dx},\ b\neq 1$
14. $\frac{d}{dx}(\log_b x)=\frac{1}{(\ln b)x},\ b\neq 1$
15. $\frac{d}{dx}(\log_b u)=\frac{1}{(\ln b)u}\cdot\frac{du}{dx},\ b\neq 1$
16. $\frac{d}{dx}(\sin x)=\cos x$
17. $\frac{d}{dx}(\cos x)=-\sin x$
18. $\frac{d}{dx}(\tan x)=\sec^2 x$
19. $\frac{d}{dx}(\cot x)=-\csc^2 x$
20. $\frac{d}{dx}(\sec x)=\sec x\tan x$
21. $\frac{d}{dx}(\csc x)=-\csc x\cot x$
22. $\frac{d}{dx}(\sin u)=\cos u\cdot\frac{du}{dx}$
23. $\frac{d}{dx}(\cos u)=-\sin u\cdot\frac{du}{dx}$
24. $\frac{d}{dx}(\tan u)=\sec^2 u\cdot\frac{du}{dx}$
25. $\frac{d}{dx}(\cot u)=-\csc^2 u\cdot\frac{du}{dx}$
26. $\frac{d}{dx}(\sec u)=(\sec u\tan u)\cdot\frac{du}{dx}$
27. $\frac{d}{dx}(\csc u)=(-\csc u\cot u)\cdot\frac{du}{dx}$
28. $\frac{d}{dx}(\sin^{-1} x)=\frac{1}{\sqrt{1-x^2}}$
29. $\frac{d}{dx}(\sin^{-1} u)=\frac{1}{\sqrt{1-u^2}}\cdot\frac{du}{dx}$
30. $\frac{d}{dx}(\cos^{-1} x)=\frac{-1}{\sqrt{1-x^2}}$
31. $\frac{d}{dx}(\cos^{-1} u)=\frac{-1}{\sqrt{1-u^2}}\cdot\frac{du}{dx}$
32. $\frac{d}{dx}(\tan^{-1} x)=\frac{1}{1+x^2}$

33. $\frac{d}{dx}(\tan^{-1} u) = \frac{1}{1+u^2} \cdot \frac{du}{dx}$

34. $\frac{d}{dx}(\cot^{-1} x) = \frac{-1}{1+x^2}$

35. $\frac{d}{dx}(\cot^{-1} u) = \frac{-1}{1+u^2} \cdot \frac{du}{dx}$

36. $\frac{d}{dx}(\sec^{-1} x) = \frac{1}{|x|\sqrt{x^2-1}}$

37. $\frac{d}{dx}(\sec^{-1} u) = \frac{1}{|u|\sqrt{u^2-1}} \cdot \frac{du}{dx}$

38. $\frac{d}{dx}(\csc^{-1} x) = \frac{-1}{|x|\sqrt{x^2-1}}$

39. $\frac{d}{dx}(\csc^{-1} u) = \frac{-1}{|u|\sqrt{u^2-1}} \cdot \frac{du}{dx}$

40. $\frac{d}{dx}(|x|) = \frac{|x|}{x}, x \neq 0$

41. $\frac{d}{dx}(|u|) = \frac{|u|}{u} \cdot \frac{du}{dx}, u \neq 0$

42. $\frac{d}{dx}(kf(x)) = k\frac{d}{dx}(f(x)) = k f'(x).$

43. $\frac{d}{dx}\left(\frac{f(x)}{k}\right) = \frac{d}{dx}\left(\frac{1}{k}f(x)\right) = \frac{1}{k}\frac{d}{dx}(f(x))$

$= \frac{1}{k}f'(x), k \neq 0$

44. $\frac{d}{dx}(f(x)+g(x)) = f'(x)+g'(x).$

45. $\frac{d}{dx}(f(x)-g(x)) = f'(x)-g'(x).$

46. $\frac{d}{dx}(f(x)g(x)) = f(x)g'(x)+g(x)f'(x)$

47. $\frac{d}{dx}\left(\frac{f(x)}{g(x)}\right) = \frac{g(x)f'(x)-f(x)g'(x)}{(g(x))^2}, g(x) \neq 0$

48. $\frac{d}{dx}[f(g(x))] = f'(g(x))g'(x)$

# APPENDIX C
# Integral formulas

1. $\int dx = x + C$

2. $\int k\,dx = kx + C$

3. $\int x^n dx = \frac{x^{n+1}}{n+1} + C,\ n \neq -1$

4. $\int x^{-1} dx = \int \frac{1}{x} dx = \ln|x| + C$

5. $\int e^x dx = e^x + C$

6. $\int e^{kx} dx = \frac{1}{k} e^{kx} + C, k \neq 0$

7. $\int b^x dx = \frac{1}{\ln b} b^x + C, b > 0, b \neq 1$

8. $\int b^{kx} dx = \frac{1}{k \ln b} b^{kx} + C, b > 0, b \neq 1, k \neq 0$

9. $\int \sin x\,dx = -\cos x + C$

10. $\int \sin(kx)\,dx = -\frac{1}{k}\cos(kx) + C, k \neq 0$

11. $\int \cos x\,dx = \sin x + C$

12. $\int \cos(kx)\,dx = \frac{1}{k}\sin(kx) + C, k \neq 0$

13. $\int \tan x\,dx = -\ln|\cos x| + C$

14. $\int \tan(kx) dx = -\frac{1}{k}\ln|\cos(kx)| + C, k \neq 0$

15. $\int \cot x\,dx = \ln|\sin x| + C$

16. $\int \cot(kx)\,dx = \frac{1}{k}\ln|\sin(kx)| + C, k \neq 0$

17. $\int \sec x\, dx = \ln|\sec x + \tan x| + C$

18. $\int \sec(kx)\, dx = \frac{1}{k}\ln|\sec(kx) + \tan(kx)| + C, k \neq 0$

19. $\int \csc x\, dx = -\ln|\csc x + \cot x| + C$

20. $\int \csc(kx)\, dx = -\frac{1}{k}\ln|\csc(kx) + \cot(kx)| + C, k \neq 0$

21. $\int \sec^2 x\, dx = \tan x + C$

22. $\int \sec^2(kx)\, dx = \frac{1}{k}\tan(kx) + C,\ k \neq 0$

23. $\int \csc^2 x\, dx = -\cot x + C$

24. $\int \csc^2(kx)\, dx = -\frac{1}{k}\cot(kx) + C, k \neq 0$

25. $\int \sec x \tan x\, dx = \sec x + C$

26. $\int \sec(kx)\tan(kx)\, dx = \frac{1}{k}\sec(kx) + C, k \neq 0$

27. $\int \csc x \cot x\, dx = -\csc x + C$

28. $\int \csc(kx)\cot(kx)\, dx = -\frac{1}{k}\csc(kx) + C, k \neq 0$

29. $\int \frac{1}{\sqrt{1-x^2}}\, dx = \sin^{-1} x + C = -\cos^{-1} x + C$

30. $\int \frac{1}{\sqrt{a^2 - x^2}}\, dx = \sin^{-1}\left(\frac{x}{a}\right) + C = -\cos^{-1}\left(\frac{x}{a}\right) + C, a > 0$

31. $\int \frac{1}{1+x^2}\, dx = \tan^{-1} x + C = -\cot^{-1} x + C$

32. $\int \frac{1}{a^2 + x^2}\, dx = \frac{1}{a}\tan^{-1}\frac{x}{a} + C = -\frac{1}{a}\cot^{-1}\left(\frac{x}{a}\right) + C, a > 0$

33. $\int \frac{1}{|x|\sqrt{x^2 - 1}}\, dx = \sec^{-1} x + C = -\csc^{-1} x + C$

34. $\int \frac{1}{|x|\sqrt{x^2 - a^2}}\, dx = \frac{1}{a}\sec^{-1}\left(\frac{x}{a}\right) + C = -\frac{1}{a}\csc^{-1}\left(\frac{x}{a}\right) + C, c > 0$

35. $\int xe^x\, dx = e^x(x-1) + C$

36. $\int x^2 e^x\, dx = e^x(x^2 - 2x + 2) + C$

37. $\int xe^{kx}\, dx = \frac{1}{k^2}e^{kx}(kx - 1) + C, k \neq 0$

38. $\int \frac{1}{1+e^x}\,dx = x - \ln(1+e^x) + C$

39. $\int \frac{1}{a+be^{kx}}\,dx = \frac{x}{a} - \frac{1}{ak}\ln(a+be^{kx}) + C,\ a, b > 0,\ k \neq 0$

40. $\int \ln x\,dx = x\ln x - x + C$

41. $\int (\ln x)^2\,dx = 2x - 2x\ln x + x(\ln x)^2 + C$

42. $\int x(\ln x)\,dx = \frac{x^2 \ln x}{2} - \frac{x^2}{4} + C$

43. $\int x^n(\ln x)\,dx = \frac{x^{n+1}\ln x}{n+1} - \frac{x^{n+1}}{(n+1)^2} + C,\ n \neq -1$

44. $\int \frac{1}{x^2-a^2}\,dx = \frac{1}{2a}\ln\left|\frac{x-a}{x+a}\right| + C,\ a \neq 0,\ x^2 > a^2$

45. $\int \frac{1}{ax+b}\,dx = \frac{1}{a}\ln|ax+b| + C,\ a \neq 0$

46. $\int \frac{1}{(ax+b)^2}\,dx = -\frac{1}{a(ax+b)} + C,\ a, b \neq 0$

47. $\int \frac{1}{x(ax+b)}\,dx = \frac{1}{b}\ln\left|\frac{x}{ax+b}\right| + C,\ a, b \neq 0$

48. $\int \frac{x}{ax+b}\,dx = \frac{x}{a} - \frac{b}{a^2}\ln|ax+b| + C$

49. $\int \frac{x}{(ax+b)^2}\,dx = \frac{1}{a^2}\left(\frac{b}{ax+b} + \ln|ax+b|\right) + C,\ a, b \neq 0$

50. $\int \frac{1}{(ax+b)(cx+d)}\,dx = \frac{1}{ad-bc}\ln\left|\frac{ax+b}{cx+d}\right| + C,\ a, b, c, d \neq 0$

51. $\int \sqrt{ax+b}\,dx = \frac{2}{3a}(ax+b)^{\frac{3}{2}} + C$

52. $\int x\sqrt{ax+b}\,dx = \frac{2(3ax-2b)}{15a^2}(ax+b)^{\frac{3}{2}} + C$

53. $\int \sqrt{x^2 \pm a^2}\,dx = \frac{x}{2}\sqrt{x^2 \pm a^2} \pm \frac{a^2}{2}\ln\left|x + \sqrt{x^2 \pm a^2}\right| + C$

54. $\int \frac{1}{\sqrt{x^2 \pm a^2}}\,dx = \ln\left|x + \sqrt{x^2 \pm a^2}\right| + C$

55. $\int \frac{1}{x\sqrt{ax+b}}\,dx = \frac{1}{\sqrt{b}}\ln\left|\frac{\sqrt{ax+b}-\sqrt{b}}{\sqrt{ax+b}+\sqrt{b}}\right| + C,\ b > 0$

56. $\int x\sin x\,dx = \sin x - x\cos x + C$

57. $\int x\cos x\,dx = \cos x + x\sin x + C$

58. $\int \frac{1}{\sin x\cos x}\,dx = \ln|\tan x| + C$

59. $\int \frac{1}{1+\sin x}\,dx = \tan x - \sec x + C$

60. $\int \frac{1}{1+\cos x}\,dx = -\cot x + \csc x + C$

61. $\int \sqrt{a^2+x^2}\,dx = \frac{x}{2}\sqrt{a^2+x^2} + \frac{a^2}{2}\ln|x+\sqrt{a^2+x^2}| + C$

62. $\int \frac{\sqrt{a^2+x^2}}{x}\,dx = \sqrt{a^2+x^2} - a\ln\left|\frac{a+\sqrt{a^2+x^2}}{x}\right| + C$

63. $\int kf(x)dx = k\int f(x)dx$

64. $\int \frac{f(x)dx}{k} = \frac{1}{k}\int f(x)dx,\ k \neq 0$

65. $\int [f(x) \pm g(x)]dx = \int f(x)dx \pm \int g(x)dx$

66. $\int f(g(x))g'(x)dx = \int f(u)du,\ u = g(x)$

67. $\int u\,dv = u\cdot v - \int v\cdot du$

# Answer key

## LIMITS

## 1 The limit concept

**1·1**

1. a. $-2.50175$ b. $-2.48259$ c. $f(x)$ is close to $-2.5$ when $x$ is close to 3.
2. a. $-0.99750$ b. $-1.00881$ c. $f(x)$ is close to $-1$ when $x$ is close to 1.
3. a. $0.003$ b. $-0.003$ c. $f(x)$ is close to 0 when $x$ is close to 0.

**1·2**

1. 5/4
2. Does not exist
3. $\sqrt{8}$
4. Approximately 58.348
5. 5/11
6. 9/11
7. 0
8. Does not exist
9. Does not exist
10. $-13/6$

## 2 Special limits

**2·1**

1. 1/7
2. $2x$
3. 6
4. 5
5. $-4/3$
6. 1/10
7. 4
8. $-4$
9. $1/2\sqrt{x}$
10. Does not exist

**2·2**

1. $\infty$
2. 0
3. $-\infty$
4. 1/18
5. 0
6. 0
7. 0
8. $\infty$
9. $-\infty$
10. 0

**2·3**

1. 5
2. 4
3. $-4$
4. $\sqrt{3}$
5. Does not exist
6. 6
7. $\infty$
8. $\infty$
9. 8
10. 8

## 3 Continuity

**3·1**

1. Not continuous
2. Continuous
3. Not continuous
4. Not continuous
5. Continuous
6. Not continuous
7. Continuous
8. Continuous
9. Continuous
10. Continuous

**3·2**

1. The tangent function is discontinuous for $3x = \frac{(2n-1)\pi}{2}$ for integers $n$, so you have continuity at $c$ when $c \neq \frac{(2n-1)\pi}{6}$.
2. The tangent and cosine functions are both continuous at 4, and so the sum is continuous at 4.
3. The cosine is not 0 at 5 and since all the functions are continuous at 5, then $f$ is continuous at 5.
4. The tangent function is not continuous at $\frac{\pi}{2}$, so $t$ is not continuous at $\frac{\pi}{2}$.
5. Since $x > 1$, the radicand is positive, and so the function $H$ is continuous at all values of $x > 1$.
6. The sine function is 0 at integral multiples of $\pi$, so the function $G$ is discontinuous at those values.
7. The sine and cosine functions are continuous on the real line, so the function $V$ is continuous there.
8. This is a disguised trig identity, so $T(x) = 1$ and therefore the function $T$ is continuous at $\frac{\pi}{11}$.
9. Since $\sin x$ is 0 at $2\pi$ and $6\pi$, $f$ is discontinuous at those points.
10. The square root function is not defined at $x = 11$, and so the function $g$ is not continuous at $x = 11$.

**3·3**

1. By inspection you can see that $f(-2)$ is positive and $f(0)$ is negative, so there is a zero between $x = -2$ and $x = 0$.
2. $g(x)$ is always positive on $[-2.5, 2]$, so there are no zeros in the interval.
3. The function is not continuous in the interval $[-5, 0]$, so the IVT does not apply.
4. The only value for which the function is 0 is $x = 0$, and this value is not in the interval $[10, 12]$.
5. The IVT does not apply in this case since $f(-2) = f(2) = 4$.
6. The function is continuous and changes sign in the interval, so there is a zero in the interval. The zero is approximately $x \approx 0.37$.
7. The function changes sign at the end points and is continuous in the interval, so there is a zero in the interval. The zero is at $x = \sqrt[3]{3} \approx 1.44$.
8. The function changes sign at the end points and is continuous in the interval, so there is a zero in the interval. The zero is $x = 0$.
9. The function is continuous and changes sign in the interval, so there is a zero in the interval. The zero is $x = \frac{5\pi}{2} \approx 7.85$.
10. The function is continuous and changes sign in the interval, so there is a zero in the interval. The zero is at $x = 0$.

# II DIFFERENTIATION

## 4 Definition of the derivative and derivatives of some simple functions

**4·1**

1. $f'(x) = 0$
2. $f'(x) = 7$
3. $f'(x) = -3$
4. $f'(x) = -3$
5. $f'(x) = -\frac{3}{4}$
6. $f'(x) = 10x + 1$

7. $f'(x)=3x^2+13$

8. $f'(x)=6x^2$

9. $f'(x)=\frac{1}{x^2}$

10. $f'(x)=-\frac{1}{2x\sqrt{x}}$

**4·2**

1. $\frac{d}{dx}(7)=0$
2. $\frac{d}{dx}(5)=0$
3. $\frac{d}{dx}(0)=0$
4. $\frac{d}{dt}(-3)=0$
5. $\frac{d}{dx}(\pi)=0$
6. $\frac{d}{dx}(25)=0$
7. $\frac{d}{dt}(100)=0$
8. $\frac{d}{dx}(2^3)=0$
9. $\frac{d}{dx}\left(-\frac{1}{2}\right)=0$
10. $\frac{d}{dx}(\sqrt{41})=0$

**4·3**

1. $f'(x)=9$
2. $g'(x)=-75$
3. $f'(x)=1$
4. $y'=50$
5. $f'(t)=2$
6. $f'(x)=\pi$
7. $f'(x)=-\frac{3}{4}$
8. $s'(t)=100$
9. $z'(x)=0.08$
10. $f'(x)=\sqrt{41}$

**4·4**

1. $f'(x)=3x^2$
2. $g'(x)=100x^{99}$
3. $f'(x)=\frac{1}{4x^{\frac{3}{4}}}$
4. $y'=\frac{1}{2x^{\frac{1}{2}}}$
5. $f'(t)=1$
6. $f'(x)=\pi x^{\pi-1}$
7. $f'(x)=-\frac{5}{x^6}$
8. $s'(t)=\frac{0.6}{t^{0.4}}$
9. $h'(s)=\frac{4}{5s^{\frac{1}{5}}}$
10. $f'(x)=-\frac{2}{3x^{\frac{5}{3}}}$

**4·5**

1. $f'(5)=75$
2. $g'(25)=0$
3. $f'(81)=\frac{1}{108}$
4. $\left.\frac{dy}{dx}\right|_{x=49}=\frac{1}{14}$
5. $f'(19)=1$
6. $f'(10)=\pi(10)^{\pi-1}\approx 435.2538$
7. $f'(2)=-\frac{5}{64}$
8. $s'(32)=0.15$
9. $h'(32)=\frac{2}{5}$
10. $\left.\frac{dy}{dx}\right|_{64}=-\frac{1}{1536}$

# 5 Rules of differentiation

**5·1**

1. $f'(x)=6x^2$
2. $g'(x)=4x^{99}$
3. $f'(x)=\frac{5}{x^{\frac{3}{4}}}$
4. $y'=-\frac{8}{x^{\frac{1}{2}}}$
5. $f'(t)=\frac{2}{3}$
6. $f'(x)=\frac{x^{\pi-1}}{2}$
7. $f'(x)=-\frac{50}{x^6}$
8. $s'(t)=\frac{60}{t^{0.4}}$
9. $h'(s)=-\frac{20}{s^{\frac{1}{5}}}$
10. $f'(x)=-\frac{1}{6x^{\frac{5}{3}}}$
11. $f'(3)=54$
12. $g'(1)=4$
13. $f'(81)=\frac{5}{27}$
14. $\left.\frac{dy}{dx}\right|_{25}=-1.6$
15. $f'(200)=\frac{2}{3}$

**5·2**

1. $f'(x)=7x^6+20x^9$
2. $h'(x)=0-10x=-10x$
3. $g'(x)=100x^{99}-200x^4$
4. $C'(x)=200-80x$
5. $y'=\frac{15}{x^2}$
6. $s'(t)=32t-\frac{2}{3}$
7. $g'(x)=4x^{99}-\frac{10}{x^{\frac{1}{2}}}$
8. $y'=\frac{2.4}{x^{0.8}}+0.45$
9. $q'(v)=\frac{2}{5v^{\frac{3}{5}}}-\frac{9}{v^{\frac{2}{5}}}$
10. $f'(x)=\frac{-5}{x^3}+5x$
11. $h'\left(\frac{1}{2}\right)=-5$
12. $C'(300)=-23{,}800$
13. $s'(0)=-\frac{2}{3}$
14. $q'(32)=-2.2$
15. $f(6)=29\frac{211}{216}$

**5·3**

1. $f'(x)=12x^2-12x+6$
2. $h'(x)=-20x^4+32x^3+60x^2-2x+2$
3. $g'(x)=\frac{15}{x^2}+3$
4. $C'(x)=1900-80x$
5. $y'=\frac{25}{2\sqrt{x}}+\frac{75}{2x\sqrt{x}}$
6. $s'(t)=40t+\frac{1}{2}$
7. $g'(x)=\frac{40x^{\frac{7}{3}}}{3}+\frac{28x^{\frac{4}{3}}}{3}$
8. $f'(x)=-\left(\frac{4}{x^3}+\frac{10}{x^6}\right)$

9. $q'(v) = \frac{70}{v^3} + 4v$

10. $f'(x) = -\frac{22x^{\frac{8}{3}}}{3} + 18x^2 - \frac{2}{x^{\frac{1}{3}}}$

11. $f'(1.5) = 15$

12. $g'(10) = 3.15$

13. $C'(150) = -10{,}100$

14. $\left.\frac{dy}{dx}\right|_{x=25} = 2.8$

15. $f'(2) = -\frac{21}{32} = -0.65625$

**5·4**

1. $f'(x) = -\frac{11}{(3x-1)^2}$

2. $h'(x) = -\frac{1}{2x^2} - \frac{5}{8}$

3. $g'(x) = -\frac{5}{2x^{\frac{3}{2}}}$

4. $f'(x) = \frac{6x\sqrt{x} + 27x + 1}{4\sqrt{x}(\sqrt{x}+3)^2}$

5. $y' = \frac{15}{x^2}$

6. $s'(t) = \frac{4t\sqrt{t} + 9t + 3}{2\sqrt{t}(2\sqrt{t}+3)^2}$

7. $g'(x) = \frac{105x^{94} + 1000x^{99}}{\left(\frac{1}{x^5} + 10\right)^2}$

8. $y' = \frac{-40x^4 + 105x^2 - 64x}{(8x^2-7)^2}$

9. $q'(v) = \frac{v^8 - 6v^3 - 4v^5 - 6}{v^4\left(v^2 - \frac{1}{v^3}\right)^2}$

10. $f'(x) = \frac{-64 - 64x^2}{x\left(\frac{4}{x^2} + 8\right)^2}$

11. $f'(25) = -\frac{11}{5476}$

12. $h'(0.2) = -13.125$

13. $g'(0.25) = -20$

14. $\left.\frac{dy}{dx}\right|_{10} = 0.15$

15. $g'(1) = \frac{1105}{121}$

**5·5**

1. $f'(x) = 18x(3x^2-10)^2$

2. $g'(x) = 720x(3x^2-10)^2$

3. $h'(x) = -\frac{180x}{(3x^2-10)^4}$

4. $h'(x) = 1 + \frac{3}{\sqrt{x}}$

5. $f'(u) = -\left(\frac{6}{u^3} + 3\right)\left(\frac{1}{u^2} - u\right)^2$

6. $y' = \frac{-6x}{(x^2-8)^4}$

7. $y' = \frac{6x^2 + 5}{2\sqrt{2x^3 + 5x + 1}}$

8. $s'(t) = \frac{6t^2 + 5}{3(2t^3 + 5t)^{\frac{2}{3}}}$

9. $f'(x) = -\frac{100}{(2x-6)^6}$

10. $C'(t) = -\frac{375}{(15t+120)^{\frac{3}{2}}}$

11. $f'(x) = 1.5138 \cdot 10^7$

12. $h'(x) = -\frac{540}{83{,}521}$

13. $f'(x) = 1\frac{1}{4}$

14. $f'(2) = -\frac{735}{64}$

15. $y' = \frac{-3}{512}$

**5·6**

1. $\frac{dy}{dx} = \frac{-2y}{x}$, provided $x \neq 0$
2. $\frac{dy}{dx} = \frac{6xy - y^3}{3xy^2 - 3x^2 - 5}$
3. $\frac{dy}{dx} = -\frac{\sqrt{y}}{\sqrt{x}}$
4. $\frac{dy}{dx} = -\frac{y^2}{x^2}$
5. $\frac{dy}{dx} = -\frac{x}{y}$
6. $\left.\frac{dy}{dx}\right|_{(3,1)} = -\frac{2}{3}$
7. $\left.\frac{dy}{dx}\right|_{(5,2)} = -\frac{13}{5}$
8. $\left.\frac{dy}{dx}\right|_{(4,9)} = -\frac{3}{2}$
9. $\left.\frac{dy}{dx}\right|_{(5,10)} = -4$
10. $\left.\frac{dy}{dx}\right|_{(2,1)} = -2$

# 6 Additional derivatives

**6·1**

1. $f'(x) = 20e^x$
2. $y' = 3e^{3x}$
3. $g'(x) = 15x^2e^{5x^3}$
4. $y' = -60x^2e^{5x^3}$
5. $h'(x) = -30x^2e^{-10x^3}$
6. $f'(x) = 30x + 10e^x$
7. $g'(x) = (7 - 6x^2)e^{7x - 2x^3}$
8. $f'(t) = 50e^{0.5t}$
9. $g'(t) = 5000e^{2t+1}$
10. $f'(x) = -\frac{x}{\sqrt{2\pi}} \cdot e^{-\frac{x^2}{2}}$

**6·2**

1. $f'(x) = \frac{20}{x}$
2. $y' = \frac{1}{x}$
3. $g'(x) = \frac{3}{x}$
4. $y' = -\frac{12}{x}$
5. $h'(x) = \frac{3}{x}$
6. $f'(x) = 30x + \frac{10}{x}$
7. $g'(x) = \frac{7 - 6x^2}{7x - 2x^3}$
8. $f'(t) = \frac{6t + 5}{3t^2 + 5t - 20}$
9. $g'(t) = 1$
10. $f'(x) = \frac{1}{x \ln x}$

**6·3**

1. $f'(x) = 20(\ln 3)(3^x)$
2. $y' = 3(\ln 5)(5^{3x})$
3. $g'(x) = \frac{d}{dx}(2^{5x^3}) = 15x^2(\ln 2)(2^{5x^3})$
4. $y' = -60x^2(\ln 2)(2^{5x^3})$
5. $h'(x) = -30x^2(\ln 4)(4^{-10x^3})$
6. $f'(x) = 30x + 30(\ln 5)(5^{3x})$

7. $g'(x) = (\ln 3)(3^{7x-2x^3})(7-6x^2)$

8. $f'(t) = 50(\ln 10)(10^{0.5t})$

9. $g'(t) = 5000(\ln 5)(5^{2t+1})$

10. $f'(x) = -x(\ln 8)\left(8^{-\frac{x^2}{2}}\right)$

**6·4**

1. $f'(x) = \dfrac{20}{x \ln 4}$

2. $y' = \dfrac{1}{x \ln 10}$

3. $g'(x) = \dfrac{3}{x \ln 8}$

4. $y' = -\dfrac{12}{x \ln 8}$

5. $h'(x) = \dfrac{3}{x \ln 5}$

6. $f'(x) = 30x + \dfrac{10}{x \ln 2}$

7. $g'(x) = \dfrac{7-6x^2}{(7x-2x^3)\ln 6}$

8. $f'(t) = \dfrac{6t+5}{(3t^2+5t-20)\ln 16}$

9. $g'(t) = \dfrac{1}{\ln 2}$

10. $f'(x) = \dfrac{1}{x(\ln 10)(\ln x)}$

**6·5**

1. $f'(x) = 15\cos 3x$

2. $h'(x) = -x\sin(2x^2)$

3. $g'(x) = 3\sec^2\left(\dfrac{3x}{5}\right)$

4. $f'(x) = 20\sec 2x \tan 2x$

5. $y' = 4x^2 \sec(2x^3)\tan(2x^3)$

6. $s'(t) = -20\csc^2 5t$

7. $g'(x) = 12\tan^2\left(\dfrac{2x}{3}\right)\sec^2\left(\dfrac{2x}{3}\right) - \dfrac{10}{\sqrt{x}}$

8. $f'(x) = 2x\cos x + 2\sin x - 2\sin 2x$

9. $h'(x) = \dfrac{3\cos 3x}{(1+\sin 3x)^2}$

10. $f'(x) = 2e^{4x}\cos 2x + 4e^{4x}\sin 2x$

**6·6**

1. $f'(x) = -\dfrac{3x^2}{\sqrt{1-x^6}}$

2. $h'(x) = -\dfrac{e^x}{\sqrt{1-e^{2x}}}$

3. $g'(x) = \dfrac{2x}{1+x^4}$

4. $f'(x) = -\dfrac{7}{1+(7x-5)^2}$

5. $y' = \dfrac{x^2}{\sqrt{1-25x^6}}$

6. $f'(x) = -\dfrac{2x}{\sqrt{1-x^4}}$

7. $h'(x) = -\dfrac{1}{|x|\sqrt{4x^2-1}}$

8. $g'(x) = \dfrac{4}{|x|\sqrt{\dfrac{x^2}{4}-1}}$

9. $f'(x) = \dfrac{14x^2}{\sqrt{1-49x^4}} + \sin^{-1}(7x^2)$

10. $y' = -\dfrac{x}{|x|\sqrt{1-x^2}}$

**6·7**

1. $f'''(x) = 210x^4 + 1440x^7$

2. $h''(x) = -\dfrac{2}{9}x^{-\frac{5}{3}} = -\dfrac{2}{9x^{\frac{5}{3}}}$

3. $g^{(5)}(x) = 0$

4. $f^{(4)}(x) = 5e^x$

5. $\dfrac{d^3y}{d^3x} = -27\cos 3x$

6. $s''(t) = 32$

7. $D_x^3[g(x)] = \dfrac{2}{x^3}$

8. $f^{(4)}(x) = \dfrac{16{,}800}{x^9}$

9. $f'''(x) = 8(\ln 3)^3(3^{2x})$

10. $\dfrac{d^4y}{d^4x} = -\dfrac{6}{x^4 \ln 2}$

# III INTEGRATION

## 7 Indefinite integral and basic integration formulas and rules

**7·1**

1. $\dfrac{d}{dx}(100x + C) = 100$
2. $\dfrac{d}{dx}(3x^2 + C) = 6x$
3. $\dfrac{d}{dx}(x^3 + 2x^2 - 5x + C) = 3x^2 + 4x - 5$
4. $\dfrac{d}{dx}\left(\dfrac{2}{7}x^{\frac{7}{2}} + \dfrac{2}{3}x^{\frac{3}{2}} + C\right) = (x^2 + 1)\sqrt{x}$
5. $\dfrac{d}{dx}\left(\dfrac{x^{e+1}}{e+1} + e^x + C\right) = x^e + e^x$
6. $\dfrac{d}{dx}\left[\dfrac{(10x+30)^4}{4} + C\right] = 10(10x+30)^3$
7. $\dfrac{d}{dx}\left[\dfrac{(x^2-3)^5}{5} + C\right] = (x^2-3)^4(2x)$
8. $\dfrac{d}{dx}\left(\dfrac{\sin^3 x}{3} + C\right) = \sin^2 x(\cos x)$
9. $\dfrac{d}{dx}\left(-\dfrac{\cos x^3}{3} + C\right) = x^2(\sin x^3)$
10. $\dfrac{d}{dx}(x\ln x - x + C) = \ln x$

**7·2**

1. $\int 8\,dx = 8x + C$
2. $\int \dfrac{3}{4}\,dx = \dfrac{3}{4}x + C$
3. $\int 9.75\,dx = 9.75x + C$
4. $\int \sqrt{3}\,dx = \sqrt{3}x + C$
5. $\int \left(\dfrac{\sqrt[3]{40}}{\sqrt{10}+15}\right) dx = \dfrac{\sqrt[3]{40}}{\sqrt{10}+15}x + C$
6. $\int 16\sqrt{2}\,dt = 16\sqrt{2}t + C$
7. $\int e^2\,dx = e^2x + C$
8. $\int 2\pi\,dr = 2\pi r + C$
9. $\int -21\,du = -21u + C$
10. $\int \dfrac{6}{e}\,dx = \dfrac{6}{e}x + C$

**7·3**

1. $\int x^5\,dx = \dfrac{x^6}{6} + C$
2. $\int \sqrt[4]{x^3}\,dx = \dfrac{4}{7}x^{\frac{7}{4}} + C$
3. $\int x^{\sqrt{2}}\,dx = \dfrac{x^{\sqrt{2}+1}}{\sqrt{2}+1} + C$
4. $\int \dfrac{1}{x^2}\,dx = -\dfrac{1}{x} + C$
5. $\int t^{100}\,dt = \dfrac{t^{101}}{101} + C$
6. $\int u^{2\pi}\,du = \dfrac{u^{2\pi+1}}{2\pi+1} + C$
7. $\int \dfrac{1}{\sqrt{x}}\,dx = 2\sqrt{x} + C$
8. $\int \dfrac{x^5}{x^2}\,dx = \dfrac{x^4}{4} + C$
9. $\int r^{-1}\,dr = \ln|r| + C$
10. $\int \dfrac{1}{t}\,dt = \ln|t| + C$

**7·4**

1. $\int e^t dt = e^t + C$
2. $\int e^{20x} dx = \frac{e^{20x}}{20} + C$
3. $\int e^{\pi x} dx = \frac{e^{\pi x}}{\pi} + C$
4. $\int e^{0.25x} dx = 4e^{0.25x} + C$
5. $\int e^{\frac{x}{5}} dx = 5e^{\frac{x}{5}} + C$
6. $\int e^{\sqrt{3}x} dx = \frac{e^{\sqrt{3}x}}{\sqrt{3}} + C$
7. $\int 4^x dx = \frac{4^x}{\ln 4} + C$
8. $\int 2^{3x} dx = \frac{2^{3x}}{3\ln 2} + C$
9. $\int 100^{0.25x} dx = \frac{4(100^{0.25x})}{\ln 100} + C$
10. $\int \pi^{\frac{x}{5}} dx = \frac{5\pi^{\frac{x}{5}}}{\ln \pi} + C$

**7·5**

1. $\int \cos v\, dv = \sin v + C$
2. $\int \sin(\tfrac{1}{2}\pi x) dx = \frac{-2\cos(\frac{1}{2}\pi x)}{\pi} + C$
3. $\int \cos(18x) dx = \frac{\sin(18x)}{18} + C$
4. $\int \sec^2(\sqrt{3}x) dx = \frac{\tan(\sqrt{3}x)}{\sqrt{3}} + C$
5. $\int \csc^2(2.5x) dx = -0.4\cot(2.5x) + C$
6. $\int \sec\left(\frac{5}{6}x\right)\tan\left(\frac{5}{6}x\right) dx = \frac{6}{5}\sec\left(\frac{5}{6}x\right) + C$
7. $\int \csc\frac{x}{3}\cot\frac{x}{3}\, dx = -3\csc\left(\frac{x}{3}\right) + C$
8. $\int \csc(ex)\cot(ex) dx = -\frac{\csc(ex)}{e} + C$
9. $\int \sin 3\theta\, d\theta = -\frac{\cos 3\theta}{3} + C$
10. $\int \cos(25\pi x) dx = \frac{\sin(25\pi x)}{25\pi} + C$

**7·6**

1. $\int \frac{1}{1+\theta^2} d\theta = \tan^{-1}\theta + C$
2. $\int \frac{dx}{\sqrt{16-x^2}} = \sin^{-1}\left(\frac{x}{4}\right) + C$
3. $\int \frac{1}{49+x^2} dx = \frac{1}{7}\tan^{-1}\left(\frac{x}{7}\right) + C$
4. $\int \frac{dt}{0.25+t^2} = 2\tan^{-1}(2t) + C$
5. $\int \frac{du}{\sqrt{u^2(u^2-1)}} = \sec^{-1} u + C$
6. $\int \frac{1}{|x|\sqrt{x^2-41}} dx = \frac{1}{\sqrt{41}}\sec^{-1}\left(\frac{x}{\sqrt{41}}\right) + C$
7. $\int \frac{1}{\sqrt{\frac{81}{100}-x^2}} dx = \sin^{-1}\left(\frac{10x}{9}\right) + C$
8. $\int \frac{1}{\pi^2+x^2} dx = \frac{1}{\pi}\tan^{-1}\left(\frac{x}{\pi}\right) + C$
9. $\int \frac{dt}{\sqrt{t^2\left(t^2-\frac{1}{4}\right)}} = 2\sec^{-1}(2t) + C$
10. $\int \frac{1}{|x|\sqrt{x^2-7}} dx = \frac{1}{\sqrt{7}}\sec^{-1}\left(\frac{x}{\sqrt{7}}\right) + C$

**7·7**

1. $\int (3x^4 - 5x^3 - 21x^2 + 36x - 10) dx = \frac{3x^5}{5} - \frac{5x^4}{4} - 7x^3 + 18x^2 - 10x + C$
2. $\int [3x^2 - 4\cos(2x)]\, dx = x^3 - 2\sin(2x) + C$
3. $\int \left(\frac{8}{t^5} + \frac{5}{t}\right) dt = -\frac{2}{t^4} + 5\ln t + C$

4. $\int\left(\frac{1}{\sqrt{25-\theta^2}}+\frac{1}{100+\theta^2}\right)d\theta=\sin^{-1}\left(\frac{\theta}{5}\right)+\frac{1}{10}\tan^{-1}\left(\frac{\theta}{10}\right)+C$

5. $\int\frac{e^{5x}-e^{4x}}{e^{2x}}dx=\frac{e^{3x}}{3}-\frac{e^{2x}}{2}+C$

6. $\int\left(\frac{x^7+x^4}{x^5}\right)dx=\frac{x^3}{3}+\ln x+C$

7. $\int\frac{1}{e^6+x^2}dx=\frac{1}{e^3}\tan^{-1}\left(\frac{x}{e^3}\right)+C$

8. $\int(x^2+4)^2dx=\int(x^4+8x^2+16)dx=\frac{x^5}{5}+\frac{8x^3}{3}+16x+C$

9. $\int\left(\frac{7}{\sqrt[3]{t}}\right)dt=\frac{21t^{\frac{2}{3}}}{2}+C$

10. $\int\frac{20+x}{\sqrt{x}}dx=40\sqrt{x}+\frac{2}{3}|x|\sqrt{x}+C$

# 8 Basic integration techniques

**8·1**

1. $\int 3(x^3-5)^4x^2dx=\frac{(x^3-5)^5}{5}+C$

2. $\int e^{x^4}x^3dx=\frac{1}{4}e^{x^4}+C$

3. $\int\frac{t}{t^2+7}dt=\frac{1}{2}\ln(t^2+7)+C$

4. $\int(x^5-3x)^{\frac{1}{4}}(5x^4-3)dx=\frac{4(x^5-3x)^{\frac{5}{4}}}{5}+C$

5. $\int\frac{x^3-2x}{(x^4-4x^2+5)^4}dx=-\frac{1}{12(x^4-4x^2+5)^3}+C$

6. $\int\frac{x^3-2x}{x^4-4x^2+5}dx=\frac{1}{4}\ln|(x^4-4x^2+5)|+C$

7. $\int x\cos(3x^2+1)dx=\frac{1}{6}\sin(3x^2+1)+C$

8. $\int\frac{3\cos^2\sqrt{x}(\sin\sqrt{x})}{\sqrt{x}}dx=-2\cos^3(\sqrt{x})+C$

9. $\int\frac{e^{2x}}{1+e^{4x}}dx=\frac{1}{2}\tan^{-1}(e^{2x})+C$

10. $\int 6t^2e^{t^3-2}dt=2e^{t^3-2}+C$

**8·2**

1. $\int 2x\sin(2x)dx=\frac{1}{2}\sin(2x)-x\cos(2x)+C$

2. $\int x^3\ln x\,dx=\frac{x^4\ln x}{4}-\frac{x^4}{16}+C$

3. $\int te^tdt=e^t(t-1)+C$

4. $\int x\cos x\,dx=x\sin x+\cos x+C$

5. $\int\cot^{-1}(x)dx=x(\cot^{-1}x)+\frac{1}{2}\ln|1+x^2|+C$

6. $\int x^2e^xdx=x^2e^x-2xe^x+2e^x+C$

7. $\int w(w-3)^2dw=\frac{w(w-3)^3}{3}-\frac{(w-3)^4}{12}+C$

8. $\int x^3\ln(4x)dx=\frac{x^4\ln(4x)}{4}-\frac{x^4}{16}+C$

9. $\int t(t+5)^{-4}dt=-\frac{t}{3(t+5)^3}-\frac{1}{6(t+5)^2}+C$

10. $\int x\sqrt{x+2}\,dx=\frac{2x(x+2)^{\frac{3}{2}}}{3}-\frac{4}{15}(x+2)^{\frac{5}{2}}+C$

**8·3**

1. $\int \cot x\,dx = \ln|\sin x| + C$
2. $\int \frac{1}{(x+2)(3x+5)}dx = -\ln\left|\frac{x+2}{3x+5}\right| + C$
3. $\int (\ln x)^2 dx = 2x - 2x\ln x + x(\ln x)^2 + C$
4. $\int x\cos x\,dx = \cos x + x\sin x + C$
5. $\int \frac{x}{(x+2)^2}dx = \frac{2}{x+2} + \ln|x+2| + C$
6. $3\int xe^x dx = 3e^x(x-1) + C$
7. $\int \sqrt{10w+3}\,dw == \frac{1}{15}(10w+3)^{\frac{3}{2}} + C$
8. $\int t(t+5)^{-1}dt = t - 5\ln|t+5| + C$
9. $\int x\sqrt{x+2}\,dx = \frac{2(3x-4)}{15}(x+2)^{\frac{3}{2}} + C$
10. $\int \frac{1}{\sin u\cos u}du = \ln|\tan u| + C$

# 9 The definite integral

**9·1**

1. $\int_{-10}^{10}(3x^2+4x-5)dx = 1900$
2. $\int_{-50}^{30} 8\,dx = 640$
3. $\int_2^7 \frac{x^5}{x^2}dx = 596.25$
4. $\int_6^{36}\frac{1}{t}dt = \ln 6$
5. $\int_{0.5\pi}^{\pi} \sec\left(\frac{5}{6}\theta\right)\tan\left(\frac{5}{6}\theta\right)d\theta = \frac{6}{5}\left(\sec\left(\frac{5\pi}{6}\right) - \sec\left(\frac{5\pi}{12}\right)\right) \approx -6.0221$
6. $\int_1^{\sqrt{3}} \frac{dx}{\sqrt{4-x^2}} = \frac{\pi}{6} \approx 0.5236$
7. $\int_1^2 (3x^4 - 5x^3 - 21x^2 + 36x - 10)dx = -\frac{103}{20} = -5.15$
8. $\int_3^5 (x^3\ln x)dx = \left(\frac{5^4\ln 5}{4} - \frac{5^4}{16}\right) - \left(\frac{3^4\ln 3}{4} - \frac{3^4}{16}\right) \approx 195.2278$
9. $\int_1^{\sqrt{3}} \cot^{-1}(x)dx = \left(\sqrt{3}\cdot\frac{\pi}{6} + \frac{1}{2}\ln(4)\right) - \left(\frac{\pi}{4} + \frac{1}{2}\ln(2)\right) \approx 0.4681$
10. $\int_2^5 \frac{1}{1+e^x}dx = 3 - \ln(1+e^5) + \ln(1+e^2) \approx 0.1202$

**9·2**

1. By Property 1, $\int_2^2 f(x)dx = 0$.
2. By Property 2, $\int_0^{-2} f(x)dx = -12$.
3. By Property 1, $\int_1^1 f(x)dx = 0$.
4. By Property 3, $\int_{-2}^2 f(x)dx = 27$.
5. By Property 4, $\int_{-2}^0 5f(x)dx = 60$.
6. By Properties 4 and 3, $\int_2^{-2} 10f(x)dx = -270$.
7. By Property 5, $\int_1^5 [f(x)+g(x)]dx = 14$.
8. By Property 5, $\int_1^5 [f(x)-g(x)]dx = -30$.
9. By Property 4, $\int_1^5 \frac{1}{2}f(x)dx = -4$.
10. By Property 4, $\int_1^5 2g(x)dx + \int_1^5 3f(x)dx = 20$.

**9·3**

1. $\frac{d}{dx}\left[\int_0^x (t^2+3)^{-5}dt\right] = \frac{1}{(x^2+3)^5}$
2. $\frac{d}{dx}\left[\int_1^x \sqrt{3t+5}\,dt\right] = \sqrt{3x+5}$
3. $\frac{d}{dx}\left[\int_\pi^{x^4} t\sin t\,dt\right] = 4x^7\sin(x^4)$
4. $\frac{d}{dx}\left[\int_{-5}^{5x^2} \sqrt[3]{t^2}\,dt\right] = 10x^2\sqrt[3]{25x}$

5. $\frac{d}{dx}\left[\int_{-10}^{x+2}(t^2-2t+1)dt\right]=x^2+2x+1$

6. $F'(x)=\sin(3x)$

7. $F'(x)=\frac{4}{4x+1}$

8. $F'(x)=6\sin^2 x\cos x$

9. $F'(x)=x^{\frac{3}{2}}$

10. $F'(x)=12x-8$

**9·4**

1. $c=0$

2. $c=\frac{16}{9}$

3. $c=2\sqrt[3]{\frac{4}{5}}$

4. $c=\sin^{-1}\left(\frac{2}{\pi}\right)$

5. $c=\frac{2}{\ln 3}$

6. $\frac{1}{(2-(-2))}\int_{-2}^{2}x^2dx=\frac{4}{3}$

7. $\frac{1}{(3-1)}\int_{1}^{3}\frac{1}{x}dx=\frac{\ln 3}{2}$

8. $\frac{1}{\left(\frac{\pi}{2}-\left(-\frac{\pi}{2}\right)\right)}\int_{-\frac{\pi}{2}}^{\frac{\pi}{2}}\cos x\,dx=\frac{2}{\pi}$

9. $\frac{1}{(4-1)}\int_{1}^{4}\frac{9}{2}\sqrt{x}\,dx=7$

10. $\frac{1}{(1-0)}\int_{0}^{1}e^x dx=e-1$

# APPLICATIONS OF THE DERIVATIVE AND THE DEFINITE INTEGRAL

## 10 Applications of the derivative

**10·1**

1. $3+\frac{1}{e}+\cos(1)$
2. 5
3. $y=-4x-8$
4. $m=3x^2-12x+9$
5. (1, –2)
6. (2, –41) and (–2, 55)
7. $y=-x+2$
8. $y=8x-5$
9. No solution
10. $y=-2x+1$

**10·2**

1. 20 acres/hour
2. –32 ft/sec
3. 23 castings/hour
4. 160 cm/sec
5. The direction of motion changes at $t=2.5$ sec. At $t=1$ both the velocity and acceleration are 0 so we cannot glean any information at that time.
6. $v=4$ ft/sec and $a=6$ ft/sec$^2$
7. 48/5 pints/lb
8. –64 ft/sec and –32 ft/sec$^2$
9. 16 ft/sec and 196 ft
10. –17,500 gal/min

**10·3**

1. $x = 3$

2. $f'(0) = \lim_{x \to 0} \frac{|x| - 0}{x - 0} = \lim_{x \to 0} \frac{|x|}{x}$. But $\lim_{x \to 0^+} \frac{|x|}{x} = \lim_{x \to 0^+} \frac{x}{x} = 1$ and $\lim_{x \to 0^-} \frac{|x|}{x} = \lim_{x \to 0^-} \frac{-x}{x} = -1$, so the $\lim_{x \to 0} \frac{|x|}{x}$ does not exist and the function is not differentiable at $x = 0$. On the other hand, $f'(x) = 1$ when $x > 0$, and $f'(x) = -1$ when $x < 0$.

3. $\lim_{x \to 2} (x-2)^{\frac{1}{3}} = 0$ and thus the function is continuous there, but $\lim_{x \to 2} \frac{(x-2)^{\frac{1}{3}} - 0}{x - 2} = \lim_{x \to 2} \frac{1}{(x-2)^{\frac{2}{3}}}$ does not exist and the function is not differentiable at $x = 2$.

4. yes

5. no

**10·4**

1. a. $x = 0, 1, 2$ and $(0,0), \left(1, \frac{1}{4}\right), (2,0)$
   b. Incr. $[0, 1]$ and $[2, \infty)$; Decr. $(-\infty, 0], [1, 2]$
   c. $f(0) = 0$, *rel* min; $f(1) = \frac{1}{4}$, *rel* max; $f(2) = 0$, *rel* min
2. a. No critical points
   b. Incr. $(0, \infty)$; Decr. nowhere
   c. No relative extrema
3. a. $x = (4k+1)\pi, x = (4k+3)\pi$ for integers $k$ and $((4k+1)\pi, 4), ((4k+3)\pi, -4)$
   b. Incr. $[(4k-1)\pi, (4k+1)\pi]$; Decr. $[(4k+1)\pi, (4k-3)\pi]$
   c. $f$ has a *rel* max of 4 at $x = (4k+1)\pi$ and a *rel* min of $-4$ at $x = (4k+3)\pi$ where $k$ is an integer.
4. a. No critical points
   b. Incr. $(-\infty, -3]$ and $[3, \infty)$
   c. No extrema since the only zeros of $f'$ occur outside the domain of $f$.
5. a. $x = 0, -2$ and $(0, 1), (-2, 5)$
   b. Incr. $[-\infty, -2]$ and $[0, \infty)$; Decr. $[-2, 0]$
   c. $f(-2) = 5$, *rel* max; $f(0) = 1$, *rel* min
6. a. $x = 4$ at which the derivative is undefined and $(4, 2)$
   b. Incr. $[-\infty, 4]$ and Decr. $[4, \infty)$
   c. $f(4) = 2$, *rel* max
7. $a = -3$ and $b = 7$
8. All the wire should be put on the square.
9. $r = .06$ or 6% interest
10. $h = 4r$

**10·5**

1. Concave up on $x < 0$ and $x > 0$. The point $(0, 2)$ is not a point of inflection.
2. $(1, -1)$ and $(2, 0)$ are points of inflection. Concave up for $x < 1$ or $x > 2$ and concave down for $1 < x < 2$.
3. No concavity and no points of inflection.
4. $(-2, -6)$ is a point of inflection and the curve is concave up for $x < -2$ and down for $x > -2$
5. No points of inflection. Concave up for $x < 2$ and down for $x > 2$.
6. $(0, 0)$ is a point of inflection. Concave up for $x > 0$ and down for $x < 0$.
7. $(0, 0)$ is a point of inflection. Concave up for $x < 0$ and down for $x > 0$.
8. $(1, -2)$ is a point of inflection. Concave up for $x < 1$ and down for $x > 1$.

9. $(\pm\sqrt{3},-44)$ are points of inflection. Concave up for $x<-\sqrt{3}$ or $x>\sqrt{3}$ and concave down for $-\sqrt{3}<x<\sqrt{3}$.
10. $a=2$, $b=-6$, $c=0$, $d=3$.

**10·6**

1. $c=1$ and $c=-1$
2. No solution since $f$ is discontinuous at 0.
3. $c=9/4$
4. No solution since $f$ is discontinuous at 1.
5. $c=\dfrac{3\pm 2\sqrt{3}}{3}$
6. No solution since $f$ is not differentiable at 0.
7. $c=\arcsin\left(-\dfrac{2}{\pi}\right)$
8. $c=\dfrac{-3\pm\sqrt{13}}{4}$
9. $c=\arccos\left(\dfrac{\pi}{2}-1\right)$
10. Let $f(x)=\ln x$ and consider the interval $\left[1,\dfrac{8}{7}\right]$. By the MVT you get $\dfrac{1}{c}=\dfrac{\ln\left(\frac{8}{7}\right)-\ln(1)}{\frac{8}{7}-1}$ or on simplification $\dfrac{1}{c}=7\ln\left(\dfrac{8}{7}\right)$ or finally that $c=\dfrac{1}{7\ln\left(\frac{8}{7}\right)}$. Also you know that $1<c<\dfrac{8}{7}$. So you get $1<\dfrac{1}{7\ln\left(\frac{8}{7}\right)}<\dfrac{8}{7}$ and solving this inequality you get $\dfrac{1}{8}<\ln\left(\dfrac{8}{7}\right)<\dfrac{1}{7}$.

# 11 Applications of the definite integral

**11·1**

1. 81 sq. units
2. 1 sq. unit
3. 44/3 sq. units
4. 1 sq. unit
5. 104/3 sq. units
6. $\dfrac{1}{2}$ sq. unit

**11·2**

1. 32/3 sq. units
2. 9/2 sq. units
3. 1/3 sq. unit
4. $\dfrac{1}{2}$ sq. unit
5. 3/2 sq. units
6. 3/2 sq. units
7. 1 sq. unit
8. 19/30 sq. unit
9. 10/3 sq. units
10. 4/3 sq. units

**11·3**

1. $\sqrt{3}-\dfrac{\ln(2-\sqrt{3})}{2}$
2. $\sqrt{97}$
3. 12
4. 17/12
5. 123/32
6. 33/16
7. $2-\sqrt{2}+\ln(1+\sqrt{2})-\dfrac{\ln 3}{2}$
8. 53/6
9. $2\sqrt{3}-\dfrac{4}{3}$
10. 52/27

# Worked solutions

## I LIMITS

### 1 The limit concept

**1·1**

1. a. −2.50175 b. −2.48259 c. $f(x)$ is close to −2.5 when $x$ is close to 3.
2. a. −0.99750 b. −1.00881 c. $f(x)$ is close to −1 when $x$ is close to 1.
3. a. .003 b. −.003 c. $f(x)$ is close to 0 when $x$ is close to 0.

**1·2**

1. $\lim\limits_{x\to 3}\dfrac{x^2-4}{x+1}=\dfrac{\lim\limits_{x\to 3}(x^2-4)}{\lim\limits_{x\to 3}(x+1)}=\dfrac{\lim\limits_{x\to 3}x^2-4}{\lim\limits_{x\to 3}x+1}=\dfrac{5}{4}$
2. $\lim\limits_{x\to 2}\dfrac{x^2-9}{x-2}=\dfrac{\lim\limits_{x\to 2}(x^2-9)}{\lim\limits_{x\to 2}(x-2)}=\dfrac{\lim\limits_{x\to 2}x^2-9}{\lim\limits_{x\to 2}x-2}$ This yields 0 in the denominator and −5 in the numerator, so the limit does not exist.
3. $\lim\limits_{x\to 1}\sqrt{x^3+7}=\sqrt{\lim\limits_{x\to 1}x^3+7}=\sqrt{8}$
4. $\lim\limits_{x\to \pi}5x^2+9=5\lim\limits_{x\to \pi}x^2+9=5\pi^2+9\approx 58.348022$
5. $\lim\limits_{x\to 0}\dfrac{5-3x}{x+11}=\dfrac{\lim\limits_{x\to 0}(5-3x)}{\lim\limits_{x\to 0}(x+11)}=\dfrac{5-3\lim\limits_{x\to 0}x}{\lim\limits_{x\to 0}x+11}=\dfrac{5}{11}$
6. $\lim\limits_{x\to 0}\dfrac{9+3x^2}{x^3+11}=\dfrac{\lim\limits_{x\to 0}(9+3x^2)}{\lim\limits_{x\to 0}(x^3+11)}=\dfrac{9+3\lim\limits_{x\to 0}x^2}{\lim\limits_{x\to 0}x^3+11}=\dfrac{9}{11}$
7. $\lim\limits_{x\to 1}\dfrac{x^2-2x+1}{x^2-1}=\lim\limits_{x\to 1}\dfrac{(x-1)^2}{(x-1)(x+1)}=\lim\limits_{x\to 1}\dfrac{(x-1)}{(x+1)}=\dfrac{\lim\limits_{x\to 1}(x-1)}{\lim\limits_{x\to 1}(x+1)}=\dfrac{\lim\limits_{x\to 1}x-1}{\lim\limits_{x\to 1}x+1}=0$
8. $\lim\limits_{x\to 4}\dfrac{6-3x}{x^2-16}=\dfrac{\lim\limits_{x\to 4}(6-3x)}{\lim\limits_{x\to 4}(x^2-16)}=\dfrac{6-3\lim\limits_{x\to 4}x}{\lim\limits_{x\to 4}x^2-16}$ This yields 0 in the denominator and a −6 in the numerator, so the limit does not exist.
9. $\lim\limits_{x\to -2}\sqrt{4x^3+11}=\sqrt{4\lim\limits_{x\to -2}x^3+11}$ The radicand is negative, so the limit does not exist.
10. $\lim\limits_{x\to -6}\dfrac{8-3x}{x-6}=\dfrac{\lim\limits_{x\to -6}(8-3x)}{\lim\limits_{x\to -6}(x-6)}=\dfrac{8-3\lim\limits_{x\to -6}x}{\lim\limits_{x\to -6}x-6}=-\dfrac{26}{12}=-\dfrac{13}{6}$

# 2 Special limits

**2·1**

1. $\lim\limits_{x\to 3}\dfrac{x-3}{x^2+x-12}=\lim\limits_{x\to 3}\dfrac{x-3}{(x-3)(x+4)}=\lim\limits_{x\to 3}\dfrac{1}{x+4}=\dfrac{1}{7}$

2. $\lim\limits_{h\to 0}\dfrac{(x+h)^2-x^2}{h}=\lim\limits_{h\to 0}\dfrac{x^2+2xh+h^2-x^2}{h}=\lim\limits_{h\to 0}\dfrac{2xh+h^2}{h}=\lim\limits_{h\to 0}(2x+h)=2x$

3. $\lim\limits_{x\to 4}\dfrac{x^3-64}{x^2-16}=\lim\limits_{x\to 4}\dfrac{(x-4)(x^2+4x+16)}{(x-4)(x+4)}=\lim\limits_{x\to 4}\dfrac{x^2+4x+16}{x+4}=\dfrac{16+16+16}{8}=6$

4. If $f(x)=5x+8$, $\lim\limits_{h\to 0}\dfrac{f(x+h)-f(x)}{h}=\lim\limits_{h\to 0}\dfrac{(5(x+h)+8)-(5x+8)}{h}=\lim\limits_{h\to 0}\dfrac{5h}{h}=\lim\limits_{h\to 0}5=5$

5. $\lim\limits_{x\to -3}\dfrac{5x+7}{x^2-3}=\dfrac{-15+7}{9-3}=-\dfrac{8}{6}=-\dfrac{4}{3}$

6. $\lim\limits_{x\to 25}\dfrac{\sqrt{x}-5}{x-25}=\lim\limits_{x\to 25}\dfrac{x-25}{(x-25)(\sqrt{x}+5)}=\lim\limits_{x\to 25}\dfrac{1}{\sqrt{x}+5}=\dfrac{1}{10}$

7. If $g(x)=x^2$, $\lim\limits_{x\to 2}\dfrac{g(x)-g(2)}{x-2}=\lim\limits_{x\to 2}\dfrac{x^2-4}{x-2}=\lim\limits_{x\to 2}\dfrac{(x-2)(x+2)}{x-2}=\lim\limits_{x\to 2}(x+2)=4$

8. $\lim\limits_{x\to 0}\dfrac{2x^2-4x}{x}=\lim\limits_{x\to 0}\dfrac{x(2x-4)}{x}=\lim\limits_{x\to 0}(2x-4)=-4$

9. $\lim\limits_{r\to 0}\dfrac{\sqrt{x+r}-\sqrt{x}}{r}=\lim\limits_{r\to 0}\dfrac{(\sqrt{x+r}-\sqrt{x})(\sqrt{x+r}+\sqrt{x})}{r(\sqrt{x+r}+\sqrt{x})}=\lim\limits_{r\to 0}\dfrac{(x+r)-x}{r(\sqrt{x+r}+\sqrt{x})}=\lim\limits_{r\to 0}\dfrac{1}{(\sqrt{x+r}+\sqrt{x})}=\dfrac{1}{2\sqrt{x}}$

10. $\lim\limits_{x\to 4}\dfrac{x^3+6}{x-4}$ This yields 0 in the denominator and 70 in the numerator, so the limit does not exist.

**2·2**

1. $\lim\limits_{x\to\infty}5x-7=\infty$

2. $\lim\limits_{x\to\infty}\dfrac{7}{x^3}=0$

3. $\lim\limits_{x\to-\infty}3x+95=-\infty$

4. $\lim\limits_{x\to\infty}\dfrac{x^3-x^2+47x+9}{18x^3+76x-11}=\lim\limits_{x\to\infty}\dfrac{1-\frac{1}{x}+\frac{47}{x^2}+\frac{9}{x^3}}{18+\frac{76}{x^2}-\frac{11}{x^3}}=\dfrac{1-0+0+0}{18+0-0}=\dfrac{1}{18}$

5. $\lim\limits_{x\to\infty}\dfrac{8}{4-x}=0$

6. $\lim\limits_{x\to\infty}\dfrac{x-2}{x^2-5x+6}=\lim\limits_{x\to\infty}\dfrac{\frac{1}{x}-\frac{2}{x^2}}{1-\frac{5}{x}+\frac{6}{x^2}}=\dfrac{0-0}{1-0+0}=0$

7. $\lim\limits_{x\to\infty}\dfrac{x^5+6x^3-7}{5x^6+6x^2-11}=\lim\limits_{x\to\infty}\dfrac{\frac{1}{x}+\frac{6}{x^3}-\frac{7}{x^6}}{5+\frac{6}{x^4}-\frac{11}{x^6}}=\dfrac{0+0-0}{5+0-0}=0$

8. $\lim\limits_{x\to-\infty}\dfrac{7x^4+6x^2-3x}{-3x^3-7x+5}=\lim\limits_{x\to-\infty}\dfrac{7x+\frac{6}{x}-\frac{3}{x^2}}{-3-\frac{7}{x^2}+\frac{5}{x^3}}=\infty$

9. $\lim_{x\to\infty}\frac{2x^3+8x-5}{-3x^2+4}=\lim_{x\to\infty}\frac{2x+\frac{8}{x}-\frac{5}{x^2}}{-3+\frac{4}{x^2}}=-\infty$

10. $\lim_{x\to-\infty}\frac{5}{x^2-4}=0$

**2·3**

1. $\lim_{x\to4^+}[x]+1=4+1=5$

2. $\lim_{x\to2^-}\frac{x^2-4}{x-2}=\lim_{x\to2^-}\frac{(x-2)(x+2)}{x-2}=\lim_{x\to2^-}(x+2)=4$

3. $\lim_{x\to8^+}\frac{4}{x-9}=-4$

4. $\lim_{x\to0^+}\sqrt{4x+3}=\sqrt{3}$

5. $\lim_{x\to5^+}[x-1]=4$ and $\lim_{x\to5^-}[x-1]=3$, so the limit does not exist

6. $\lim_{x\to3^-}\frac{x^2-9}{x-3}=\lim_{x\to3^-}\frac{(x-3)(x+3)}{x-3}=\lim_{x\to3^-}(x+3)=6$

7. $\lim_{x\to4^+}\frac{7}{x-4}=\infty$

8. $\lim_{x\to-4^-}\frac{x^5+x^4-8}{x+4}=\infty$

9. $\lim_{x\to4^+}\frac{x^2-16}{x-4}=\lim_{x\to4^+}\frac{(x-4)(x+4)}{x-4}=\lim_{x\to4^+}(x+4)=8$

10. $\lim_{x\to4^-}\frac{x^2-16}{x-4}=\lim_{x\to4^-}\frac{(x-4)(x+4)}{x-4}=\lim_{x\to4^-}(x+4)=8$

# 3 Continuity

**3·1**

1. Looking at $\lim_{x\to1}\sqrt{5x-7}$, when $x$ assumes values close to 1, the radicand is negative; and thus the limit does not exist, so the function is not continuous at 1.

2. $\lim_{x\to0}\frac{x^3-8}{2-x}=\lim_{x\to0}\frac{(x-2)(x^2+2x+4)}{-(x-2)}=\lim_{x\to0}-(x^2+2x+4)=-4$. Since the limit exists and equals $f(0)=\frac{0^3-8}{2-0}=-4$, the function is continuous at 0.

3. Looking at $\lim_{x\to1}\frac{4}{\sqrt{2x-3}}$, when $x$ assumes values close to 1, the radicand is negative; and thus the limit does not exist, so the function is not continuous at 1.

4. Since $\lim_{x\to3^+}[x]=3$ and $\lim_{x\to3^-}[x]=2$, the limit does not exist at 3; so the function is not continuous at 3.

5. $\lim_{x\to4}\frac{x^2+6}{x-5}=\frac{(\lim_{x\to4}x)^2+6}{\lim_{x\to4}x-5}=\frac{16+6}{4-5}=-22$, so the function is continuous at 4.

6. Looking at $\lim_{x\to3}\frac{\sqrt{x-5}}{x-5}$, when $x$ assumes values close to 3, the radicand is negative; thus, the limit does not exist, so the function is not continuous at 3.

7. $\lim_{x\to 8}\frac{\sqrt{x}-5}{x+2}=\frac{\sqrt{\lim_{x\to 8}x}-5}{\lim_{x\to 8}x+2}=\frac{\sqrt{8}-5}{10}$, so the function is continuous at 8.

8. $\lim_{x\to 5}(5x^2-\sqrt{x}+7)=\left(5(\lim_{x\to 5}x)^2-\sqrt{\lim_{x\to 5}x}+7\right)=125-\sqrt{5}+7=132-\sqrt{5}$, so the function is continuous at 5.

9. $\lim_{x\to 6}\frac{x-6}{x-2}=\frac{\lim_{x\to 6}x-6}{\lim_{x\to 6}x-2}=\frac{6-6}{4}=0$, so the function is continuous at 6.

10. $\lim_{x\to a}\frac{(x-a)^2+x-6}{x-a+3}=\frac{(\lim_{x\to a}x-a)^2+\lim_{x\to a}x-6}{\lim_{x\to a}x-a+3}=\frac{a-6}{3}$, so the function is continuous at $a$.

**3·2**

1. The tangent function is discontinuous for $3x=\frac{(2n-1)\pi}{2}$ for integers $n$, so you have continuity at $c$ when $c\neq\frac{(2n-1)\pi}{6}$.
2. The tangent and cosine functions are both continuous at 4, and so the sum is continuous at 4.
3. The cosine is not 0 at 5 and since all the functions are continuous at 5, then $f$ is continuous at 5.
4. The tangent function is not continuous at $\frac{\pi}{2}$, so $t$ is not continuous at $\frac{\pi}{2}$.
5. Since $x > 1$, the radicand is positive, and so the function $H$ is continuous at all values of $x > 1$.
6. The sine function is 0 at integral multiples of $\pi$, so the function $G$ is discontinuous at those values.
7. The sine and cosine functions are continuous on the real line, so the function $V$ is continuous there.
8. This is a disguised trig identity, so $T(x) = 1$ and therefore the function $T$ is continuous at $\frac{\pi}{11}$.
9. Since $\sin x$ is 0 at $2\pi$ and $6\pi$, $f$ is discontinuous at those points. However, $f$ has removable discontinuities at both points, and thus $f$ redefined at those points is continuous at both points.
10. The square root function is not defined at $x = 11$, and so the function $g$ is not continuous at $x = 11$.

**3·3**

1. By inspection you can see that $f(-2) = 79$ is positive and $f(0) = -5$ is negative, so there is a zero between $x = -2$ and $x = 0$.
2. $g(x)$ is always positive on $[-2.5, 2]$, so there are no zeros in the interval.
3. The function is not continuous in the interval $[-5, 0]$, so the IVT does not apply.
4. The only value for which the function is 0 is $x = 0$, and this value is not in the interval $[10, 12]$.
5. The IVT does not apply in this case since $f(-2) = f(2) = 4$.
6. The function is continuous and changes sign in the interval, so there is a zero in the interval. The zero is approximately $x \approx 0.37$.
7. The function changes sign at the end points and is continuous in the interval, so there is a zero in the interval. The zero is at $x=\sqrt[3]{3}\approx 1.44$.
8. The function changes sign at the end points and is continuous in the interval, so there is a zero in the interval. The zero is $x = 0$.
9. The function is continuous and changes sign in the interval, so there is a zero in the interval. The zero is $x=\frac{5\pi}{2}\approx 7.85$
10. The function is continuous and changes sign in the interval, so there is a zero in the interval. The zero is at $x = 0$.

# II DIFFERENTIATION

## 4 Definition of the derivative and derivatives of some simple functions

**4·1**

1. By definition, $f'(x)=\lim_{h\to 0}\frac{f(x+h)-f(x)}{h}$. Since $f(x)=4$ for all values of $x$, you have $f'(x)=$ $\lim_{h\to 0}\frac{f(x+h)-f(x)}{h}=\lim_{h\to 0}\frac{4-4}{h}=\lim_{h\to 0}\frac{0}{h}=\lim_{h\to 0}0=0.$

2. By definition, $f'(x)=\lim_{h\to 0}\frac{f(x+h)-f(x)}{h}=\lim_{h\to 0}\frac{(7(x+h)+2)-(7x+2)}{h}=\lim_{h\to 0}\frac{(7x+7h+2)-(7x+2)}{h}=$ $\lim_{h\to 0}\frac{7x+7h+2-7x-2}{h}=\lim_{h\to 0}\frac{7h}{h}=\lim_{h\to 0}7=7.$

3. By definition, $f'(x)=\lim_{h\to 0}\frac{f(x+h)-f(x)}{h}=\lim_{h\to 0}\frac{(-3(x+h)-9)-(-3x-9)}{h}=\lim_{h\to 0}\frac{(-3x-3h-9)-(-3x-9)}{h}=$ $\lim_{h\to 0}\frac{-3x-3h-9+3x+9}{h}=\lim_{h\to 0}\frac{-3h}{h}=\lim_{h\to 0}-3=-3.$

4. By definition, $f'(x)=\lim_{h\to 0}\frac{f(x+h)-f(x)}{h}=\lim_{h\to 0}\frac{(10-3(x+h))-(10-3x)}{h}=\lim_{h\to 0}\frac{(10-3x-3h)-(10-3x)}{h}=$ $\lim_{h\to 0}\frac{10-3x-3h-10+3x}{h}=\lim_{h\to 0}\frac{-3h}{h}=\lim_{h\to 0}-3=-3.$

5. By definition, $f'(x)=\lim_{h\to 0}\frac{f(x+h)-f(x)}{h}=\lim_{h\to 0}\frac{\left(-\frac{3}{4}(x+h)\right)-\left(-\frac{3}{4}x\right)}{h}=\lim_{h\to 0}\frac{-\frac{3}{4}x-\frac{3}{4}h+\frac{3}{4}x}{h}=$ $\lim_{h\to 0}\frac{-\frac{3}{4}h}{h}=\lim_{h\to 0}-\frac{3}{4}=-\frac{3}{4}.$

6. By definition, $f'(x)=\lim_{h\to 0}\frac{f(x+h)-f(x)}{h}=\lim_{h\to 0}\frac{(5(x+h)^2+(x+h)-3)-(5x^2+x-3)}{h}=$ $\lim_{h\to 0}\frac{(5(x^2+2xh+h^2)+(x+h)-3)-(5x^2+x-3)}{h}=\lim_{h\to 0}\frac{(5x^2+10xh+5h^2+x+h-3)-(5x^2+x-3)}{h}=$ $\lim_{h\to 0}\frac{5x^2+10xh+5h^2+x+h-3-5x^2-x+3}{h}=\lim_{h\to 0}\frac{10xh+5h^2+h}{h}=\lim_{h\to 0}(10x+5h+1)=10x+1.$

7. By definition, $f'(x)=\lim_{h\to 0}\frac{f(x+h)-f(x)}{h}=\lim_{h\to 0}\frac{((x+h)^3+13(x+h))-(x^3+13x)}{h}=$ $\lim_{h\to 0}\frac{((x^3+3x^2h+3xh^2+h^3)+(13x+13h))-(x^3+13x)}{h}=\lim_{h\to 0}\frac{x^3+3x^2h+3xh^2+h^3+13x+13h-x^3-13x}{h}=$ $\lim_{h\to 0}\frac{3x^2h+3xh^2+h^3+13h}{h}=\lim_{h\to 0}(3x^2+3xh+h^2+13)=3x^2+13.$

8. By definition, $f'(x)=\lim_{h\to 0}\frac{f(x+h)-f(x)}{h}=\lim_{h\to 0}\frac{(2(x+h)^3+15)-(2x^3+15)}{h}=$ $\lim_{h\to 0}\frac{(2(x^3+3x^2h+3xh^2+h^3)+15)-(2x^3+15)}{h}=\lim_{h\to 0}\frac{(2x^3+6x^2h+6xh^2+2h^3+15)-(2x^3+15)}{h}=$ $\lim_{h\to 0}\frac{2x^3+6x^2h+6xh^2+2h^3+15-2x^3-15}{h}=\lim_{h\to 0}(6x^2+6xh+2h^2)=6x^2.$

9. By definition, $f'(x)=\lim_{h\to 0}\frac{f(x+h)-f(x)}{h}=\lim_{h\to 0}\frac{\left(-\frac{1}{x+h}\right)-\left(-\frac{1}{x}\right)}{h}=\lim_{h\to 0}\frac{-\frac{1}{x+h}+\frac{1}{x}}{h}=$

$\lim_{h\to 0}\frac{-x+x+h}{hx(x+h)}=\lim_{h\to 0}\frac{h}{hx(x+h)}=\lim_{h\to 0}\frac{1}{x(x+h)}=\frac{1}{x^2}.$

10. By definition, $f'(x)=\lim_{h\to 0}\frac{f(x+h)-f(x)}{h}=\lim_{h\to 0}\frac{\left(\frac{1}{\sqrt{x+h}}\right)-\left(\frac{1}{\sqrt{x}}\right)}{h}=\lim_{h\to 0}\frac{\frac{\sqrt{x}-\sqrt{x+h}}{\sqrt{x}\sqrt{x+h}}}{h}=$

$\lim_{h\to 0}\frac{\left(\frac{\sqrt{x}-\sqrt{x+h}}{\sqrt{x}\sqrt{x+h}}\right)\left(\frac{\sqrt{x}+\sqrt{x+h}}{\sqrt{x}+\sqrt{x+h}}\right)}{h}=\lim_{h\to 0}\frac{\frac{x-(x+h)}{\sqrt{x}\sqrt{x+h}(\sqrt{x}+\sqrt{x+h})}}{h}=\lim_{h\to 0}\frac{\frac{x-x-h}{x\sqrt{x+h}+(x+h)\sqrt{x}}}{h}=$

$\lim_{h\to 0}\frac{-1}{x\sqrt{x+h}+(x+h)\sqrt{x}}=-\frac{1}{2x\sqrt{x}}.$

**4·2**

1. $\frac{d}{dx}(7)=0$
2. $\frac{d}{dx}(5)=0$
3. $\frac{d}{dx}(0)=0$
4. $\frac{d}{dt}(-3)=0$
5. $\frac{d}{dx}(\pi)=0$
6. $\frac{d}{dx}(25)=0$
7. $\frac{d}{dt}(100)=0$
8. $\frac{d}{dx}(2^3)=0$
9. $\frac{d}{dx}\left(-\frac{1}{2}\right)=0$
10. $\frac{d}{dx}(\sqrt{41})=0$

**4·3**

1. $f'(x)=9$
2. $g'(x)=-75$
3. $f'(x)=1$
4. $y'=50$
5. $f'(t)=2$
6. $f'(x)=\pi$
7. $f'(x)=-\frac{3}{4}$
8. $s'(t)=100$
9. $z'(x)=0.08$
10. $f'(x)=\sqrt{41}$

**4·4**

1. $f'(x)=3x^2$
2. $g'(x)=100x^{99}$
3. $f'(x)=\frac{1}{4}x^{-\frac{3}{4}}=\frac{1}{4x^{\frac{3}{4}}}$
4. $y'=\frac{d}{dx}(\sqrt{x})=\frac{d}{dx}\left(x^{\frac{1}{2}}\right)=\frac{1}{2}x^{-\frac{1}{2}}=\frac{1}{2x^{\frac{1}{2}}}$
5. $f'(t)=1$
6. $f'(x)=\pi x^{\pi-1}$
7. $f'(x)=\frac{d}{dx}\left(\frac{1}{x^5}\right)=\frac{d}{dx}(x^{-5})=-5x^{-6}=-\frac{5}{x^6}$
8. $s'(t)=0.6t^{-0.4}=0.6\frac{1}{t^{0.4}}=\frac{0.6}{t^{0.4}}$
9. $h'(s)=\frac{4}{5}\left(s^{-\frac{1}{5}}\right)=\frac{4}{5s^{\frac{1}{5}}}$
10. $f'(x)=\frac{d}{dx}\left(\frac{1}{\sqrt[3]{x^2}}\right)=\frac{d}{dx}\left(x^{-\frac{2}{3}}\right)=-\frac{2}{3}\left(x^{-\frac{5}{3}}\right)=-\frac{2}{3x^{\frac{5}{3}}}$

**4·5**

1. For $f(x)=x^3$, $f'(x)=3x^2$; thus, $f'(5)=3\cdot 5^2=75$.

2. For $g(x)=-100, g'(x)=0$ for all values of $x$; thus, $g'(25)=0$.

3. For $f(x)=x^{\frac{1}{4}}, f'(x)=\frac{1}{4}x^{-\frac{3}{4}}=\frac{1}{4x^{\frac{3}{4}}}$; thus, $f'(81)=\frac{1}{4(81)^{\frac{3}{4}}}=\frac{1}{4(27)}=\frac{1}{108}$.

4. For $y=\sqrt{x}, y'=\frac{d}{dx}(\sqrt{x})=\frac{d}{dx}\left(x^{\frac{1}{2}}\right)=\frac{1}{2}x^{-\frac{1}{2}}=\frac{1}{2x^{\frac{1}{2}}}$; thus, $\left.\frac{dy}{dx}\right|_{x=49}=\frac{1}{2(49)^{\frac{1}{2}}}=\frac{1}{14}$.

5. For $f(t)=t, f'(t)=1$ for all values of $t$; thus, $f'(19)=1$.

6. For $f(x)=x^{\pi}, f'(x)=\pi x^{\pi-1}$ thus, $f'(10)=\pi(10)^{\pi-1}\approx 435.2538$.

7. For $f(x)=\frac{1}{x^5}, f'(x)=\frac{d}{dx}\left(\frac{1}{x^5}\right)=\frac{d}{dx}(x^{-5})=-\frac{5}{x^6}$; thus, $f'(2)=-\frac{5}{2^6}=-\frac{5}{64}$.

8. For $s(t)=t^{0.6}, s'(t)=0.6t^{-0.4}=0.6\frac{1}{t^{0.4}}=\frac{0.6}{t^{0.4}}$; thus, $s'(32)=\frac{0.6}{32^{0.4}}=\frac{0.6}{4}=0.15$.

9. For $h(s)=s^{\frac{4}{5}}, h'(s)=\frac{4}{5}\left(s^{-\frac{1}{5}}\right)=\frac{4}{5s^{\frac{1}{5}}}$; thus, $h'(32)=\frac{4}{5(32)^{\frac{1}{5}}}=\frac{2}{5}$.

10. For $y=\frac{1}{\sqrt[3]{x^2}}, y'=\frac{d}{dx}\left(\frac{1}{\sqrt[3]{x^2}}\right)=\frac{d}{dx}\left(x^{-\frac{2}{3}}\right)=-\frac{2}{3x^{\frac{5}{3}}}$; thus, $\left.\frac{dy}{dx}\right|_{64}=-\frac{2}{3(1024)}=-\frac{1}{1536}$.

# 5 Rules of differentiation

**5·1**

1. If $f(x)=2x^3$, then $f'(x)=2\frac{d}{dx}(x^3)=2(3x^2)=6x^2$.

2. If $g(x)=\frac{x^{100}}{25}$, then $g'(x)=\frac{1}{25}\cdot\frac{d}{dx}(x^{100})=\frac{1}{25}\cdot(100x^{99})=4x^{99}$.

3. If $f(x)=20x^{\frac{1}{4}}$, then $f'(x)=20\cdot\frac{d}{dx}\left(x^{\frac{1}{4}}\right)=20\cdot\frac{1}{4x^{\frac{3}{4}}}=\frac{5}{x^{\frac{3}{4}}}$.

4. If $y=-16\sqrt{x}$, then $y'=-16\cdot\frac{d}{dx}(\sqrt{x})=-16\cdot\frac{d}{dx}\left(x^{\frac{1}{2}}\right)=-16\left(\frac{1}{2x^{\frac{1}{2}}}\right)=-\frac{8}{x^{\frac{1}{2}}}$.

5. If $f(t)=\frac{2t}{3}$, then $f'(t)=\frac{2}{3}\cdot\frac{d}{dt}(t)=\frac{2}{3}\cdot 1=\frac{2}{3}$.

6. If $f(x)=\frac{x^{\pi}}{2\pi}$, then $f'(x)=\frac{1}{2\pi}\cdot\frac{d}{dx}(x^{\pi})=\frac{1}{2\pi}\cdot\pi x^{\pi-1}=\frac{x^{\pi-1}}{2}$.

7. If $f(x)=\frac{10}{x^5}$, then $f'(x)=10\cdot\frac{d}{dx}\left(\frac{1}{x^5}\right)=10\left(-\frac{5}{x^6}\right)=-\frac{50}{x^6}$.

8. If $s(t)=100t^{0.6}$, then $s'(t)=100\cdot\frac{d}{dt}(t^{0.6})=100\cdot\frac{0.6}{t^{0.4}}=\frac{60}{t^{0.4}}$.

9. If $h(s)=-25s^{\frac{4}{5}}$, then $h'(s)=-25\cdot\frac{d}{dx}\left(s^{\frac{4}{5}}\right)=-25\cdot\frac{4}{5s^{\frac{1}{5}}}=-\frac{20}{s^{\frac{1}{5}}}$.

10. If $f(x)=\frac{1}{4\sqrt[3]{x^2}}$, then $f'(x)=\frac{1}{4}\cdot\frac{d}{dx}\left(\frac{1}{\sqrt[3]{x^2}}\right)=\frac{1}{4}\left(-\frac{2}{3x^{\frac{5}{3}}}\right)=-\frac{1}{6x^{\frac{5}{3}}}$.

11. When $f(x)=2x^3, f'(x)=6x^2$; thus, $f'(3)=6\cdot 3^2=54$.

12. When $g(x)=\frac{x^{100}}{25}, g'(x)=4x^{99}$; thus, $g'(1)=4\cdot 1^{99}=4$.

13. When $f(x)=20x^{\frac{1}{4}}, f'(x)=\frac{5}{x^{\frac{3}{4}}}$; thus, $f'(81)=\frac{5}{(81)^{\frac{3}{4}}}=\frac{5}{27}$.

14. When $y=-16\sqrt{x}, y'=-\frac{8}{x^{\frac{1}{2}}}$; thus, $\left.\frac{dy}{dx}\right|_{25}=-\frac{8}{25^{\frac{1}{2}}}=-\frac{8}{5}=-1.6$.

15. when $f(t)=\frac{2t}{3}, f'(t)=\frac{2}{3}$ for all values of $t$; thus, $f'(200)=\frac{2}{3}$.

**5·2**

1. When $f(x) = x^7 + 2x^{10}$, $f'(x)=7x^6+20x^9$.

2. When $h(x)=30-5x^2, h'(x)=0-10x=-10x$.

3. When $g(x)=x^{100}-40x^5, g'(x)=100x^{99}-200x^4$.

4. When $C(x)=1000+200x-40x^2, C'(x)=0+200-80x=200-80x$.

5. When $y=\frac{-15}{x}+25, y'=-15\frac{d}{dx}(x^{-1})+\frac{d}{dx}(25)=-15(-x^{-2})+0=15x^{-2}=\frac{15}{x^2}$.

6. When $s(t)=16t^2-\frac{2t}{3}+10, s'(t)=32t-\frac{2}{3}+0=32t-\frac{2}{3}$.

7. When $g(x)=\frac{x^{100}}{25}-20\sqrt{x}, g'(x)=4x^{99}-20\frac{d}{dx}\left(x^{\frac{1}{2}}\right)=4x^{99}-10x^{-\frac{1}{2}}=4x^{99}-\frac{10}{x^{\frac{1}{2}}}$.

8. When $y=12x^{0.2}+0.45x, y'=2.4x^{-0.8}+0.45=\frac{2.4}{x^{0.8}}+0.45$.

9. When $q(v)=v^{\frac{2}{5}}+7-15v^{\frac{3}{5}}, q'(v)=\frac{2}{5}v^{-\frac{3}{5}}+0-9v^{-\frac{2}{5}}=\frac{2}{5v^{\frac{3}{5}}}-\frac{9}{v^{\frac{2}{5}}}$.

10. When $f(x)=\frac{5}{2x^2}+\frac{5}{2x^{-2}}-\frac{5}{2}, f'(x)=\frac{5}{2}\frac{d}{dx}(x^{-2})+\frac{5}{2}\frac{d}{dx}(x^2)-\frac{5}{2}\frac{d}{dx}(1)=-5(x^{-3})+5x-0=\frac{-5}{x^3}+5x$.

11. When $h(x)=30-5x^2, h'(x)=-10x$; thus, $h'\left(\frac{1}{2}\right)=-10\left(\frac{1}{2}\right)=-5$.

12. When $C(x)=1000+200x-40x^2, C'(x)=200-80x$; thus, $C'(300)=200-80(300)=-23{,}800$.

13. When $s(t)=16t^2-\frac{2t}{3}+10, s'(t)=32t-\frac{2}{3}$; thus, $s'(0)=32(0)-\frac{2}{3}=-\frac{2}{3}$.

14. When $q(v)=v^{\frac{2}{5}}+7-15v^{\frac{3}{5}}, q'(v)=\frac{2}{5v^{\frac{3}{5}}}-\frac{9}{v^{\frac{2}{5}}}$; thus, $q'(32)=\frac{2}{5(32)^{\frac{3}{5}}}-\frac{9}{(32)^{\frac{2}{5}}}=\frac{2}{5(8)}-\frac{9}{(4)}=-2.2$.

15. When $f(x)=\frac{5}{2x^2}+\frac{5}{2x^{-2}}-\frac{5}{2}, f'(x)=\frac{-5}{x^3}+5x$; thus, $f'(6)=\frac{-5}{6^3}+5(6)=\frac{-5}{216}+30=29\frac{211}{216}$.

**5·3**

1. If $f(x)=(2x^2+3)(2x-3)$, then $f'(x)=(2x^2+3)\frac{d}{dx}(2x-3)+(2x-3)\frac{d}{dx}(2x^2+3)=(2x^2+3)(2)+$ $(2x-3)(4x)=12x^2-12x+6$.

2. If $h(x)=(4x^3+1)(-x^2+2x+5)$, then $h'(x)=(4x^3+1)\frac{d}{dx}(-x^2+2x+5)+(-x^2+2x+5)\frac{d}{dx}(4x^3+1)=$ $(4x^3+1)(-2x+2)+(-x^2+2x+5)(12x^2)=-20x^4+32x^3+60x^2-2x+2$.

3. If $g(x)=(x^2-5)\left(\frac{3}{x}\right)$, then $g'(x)=(x^2-5)\frac{d}{dx}\left(\frac{3}{x}\right)+\left(\frac{3}{x}\right)\frac{d}{dx}(x^2-5)=(x^2-5)\left(-\frac{3}{x^2}\right)+\left(\frac{3}{x}\right)(2x)=\frac{15}{x^2}+3$.

4. If $C(x)=(50+20x)(100-2x)$, then $C'(x)=(50+20x)\frac{d}{dx}(100-2x)+(100-2x)\frac{d}{dx}(50+20x)=$ $(50+20x)(-2)+(100-2x)(20)=1900-80x$.

5. If $y=\left(\frac{-15}{\sqrt{x}}+25\right)(\sqrt{x}+5)$, then $y'=\left(\frac{-15}{\sqrt{x}}+25\right)\frac{d}{dx}(\sqrt{x}+5)+(\sqrt{x}+5)\frac{d}{dx}\left(\frac{-15}{\sqrt{x}}+25\right)=$

$\left(\frac{-15}{\sqrt{x}}+25\right)\left(\frac{1}{2x^{\frac{1}{2}}}\right)+(\sqrt{x}+5)\left(\frac{15}{2x\sqrt{x}}\right)=\frac{25}{2\sqrt{x}}+\frac{75}{2x\sqrt{x}}$.

6. If $s(t)=\left(4t-\frac{1}{2}\right)\left(5t+\frac{3}{4}\right)$, then $s'(t)=\left(4t-\frac{1}{2}\right)\frac{d}{dx}\left(5t+\frac{3}{4}\right)+\left(5t+\frac{3}{4}\right)\frac{d}{dx}\left(4t-\frac{1}{2}\right)=$

$\left(4t-\frac{1}{2}\right)(5)+\left(5t+\frac{3}{4}\right)(4)=40t+\frac{1}{2}$.

7. If $g(x)=(2x^3+2x^2)(2\sqrt[3]{x})$, then $g'(x)=(2x^3+2x^2)\frac{d}{dx}\left(2x^{\frac{1}{3}}\right)+\left(2x^{\frac{1}{3}}\right)\frac{d}{dx}(2x^3+2x^2)=$

$(2x^3+2x^2)\left(\frac{2}{3}x^{-\frac{2}{3}}\right)+\left(2x^{\frac{1}{3}}\right)(6x^2+4x)=\frac{40x^{\frac{7}{3}}}{3}+\frac{28x^{\frac{4}{3}}}{3}$.

8. If $f(x)=\frac{10}{x^5}\cdot\frac{x^3+1}{5}$, then $f'(x)=(10x^{-5})\frac{1}{5}\cdot\frac{d}{dx}(x^3+1)+\left(\frac{x^3+1}{5}\right)\left(10\frac{d}{dx}x^{-5}\right)=(2x^{-5})(3x^2)+$

$\left(\frac{x^3+1}{5}\right)(-50x^{-6})=-\left(\frac{4}{x^3}+\frac{10}{x^6}\right)$.

9. If $q(v)=(v^2+7)(-5v^{-2}+2)$, then $q'(v)=(v^2+7)\frac{d}{dv}(-5v^{-2}+2)+(-5v^{-2}+2)\frac{d}{dv}(v^2+7)=$

$(v^2+7)(10v^{-3})+(-5v^{-2}+2)(2v)=\frac{70}{v^3}+4v$.

10. If $f(x)=(2x^3+3)(3-\sqrt[3]{x^2})$, then $f'(x)=(2x^3+3)\frac{d}{dx}\left(3-x^{\frac{2}{3}}\right)+\left(3-x^{\frac{2}{3}}\right)\frac{d}{dx}(2x^3+3)=$

$(2x^3+3)\left(-\frac{2}{3}x^{-\frac{1}{3}}\right)+\left(3-x^{\frac{2}{3}}\right)(6x^2)=-\frac{22x^{\frac{8}{3}}}{3}+18x^2-\frac{2}{x^{\frac{1}{3}}}$.

11. When $f(x)=(2x^2+3)(2x-3)$, $f'(x)=12x^2-12x+6$; thus, $f'(1.5)=12(1.5)^2-12(1.5)+6=15$.

12. When $g(x)=(x^2-5)\left(\frac{3}{x}\right)$, $g'(x)=\frac{15}{x^2}+3$; thus, $g'(10)=\frac{15}{10^2}+3=3.15$.

13. When $C(x)=(50+20x)(100-2x)$, $C'(x)=1900-80x$; thus, $C'(150)=1900-80(150)=-10{,}100$.

14. When $y=\left(\frac{-15}{\sqrt{x}}+25\right)(\sqrt{x}+5)$, $y'=\frac{25}{2\sqrt{x}}+\frac{75}{2x\sqrt{x}}$; thus, $\left.\frac{dy}{dx}\right|_{x=25}=\frac{25}{2\sqrt{25}}+\frac{75}{2(25)\sqrt{25}}=2.8$.

15. When $f(x)=\frac{10}{x^5}\cdot\frac{x^3+1}{5}$, $f'(x)=-\left(\frac{4}{x^3}+\frac{10}{x^6}\right)$; thus, $f'(2)=-\left(\frac{4}{2^3}+\frac{10}{2^6}\right)=-\frac{21}{32}=-0.65625$.

**5·4**

1. If $f(x)=\frac{5x+2}{3x-1}$, then $f'(x)=\frac{(3x-1)\frac{d}{dx}(5x+2)-(5x+2)\frac{d}{dx}(3x-1)}{(3x-1)^2}=\frac{(3x-1)(5)-(5x+2)(3)}{(3x-1)^2}=$

$-\frac{11}{(3x-1)^2}$.

2. If $h(x)=\frac{4-5x^2}{8x}$, then $h'(x)=\frac{(8x)\frac{d}{dx}(4-5x^2)-(4-5x^2)\frac{d}{dx}(8x)}{(8x)^2}=\frac{(8x)(-10x)-(4-5x^2)(8)}{(8x)^2}=$

$-\frac{1}{2x^2}-\frac{5}{8}$.

3. If $g(x)=\frac{5}{\sqrt{x}}$, then $g'(x)=\frac{\left(x^{\frac{1}{2}}\right)\frac{d}{dx}(5)-(5)\frac{d}{dx}\left(x^{\frac{1}{2}}\right)}{\left(x^{\frac{1}{2}}\right)^2}=\frac{\left(x^{\frac{1}{2}}\right)(0)-(5)\left(\frac{1}{2}x^{-\frac{1}{2}}\right)}{x}=-\frac{5}{2x^{\frac{3}{2}}}$.

4. If $f(x)=\dfrac{3x^{\frac{3}{2}}-1}{2x^{\frac{1}{2}}+6}$, then $f'(x)=\dfrac{\left(2x^{\frac{1}{2}}+6\right)\dfrac{d}{dx}\left(3x^{\frac{3}{2}}-1\right)-\left(3x^{\frac{3}{2}}-1\right)\dfrac{d}{dx}\left(2x^{\frac{1}{2}}+6\right)}{\left(2x^{\frac{1}{2}}+6\right)^2}=$

$$\frac{\left(2x^{\frac{1}{2}}+6\right)\left(\frac{9}{2}x^{\frac{1}{2}}\right)-\left(3x^{\frac{3}{2}}-1\right)\left(x^{-\frac{1}{2}}\right)}{\left(2x^{\frac{1}{2}}+6\right)^2}=\frac{6x+27x^{\frac{1}{2}}+x^{-\frac{1}{2}}}{\left(2x^{\frac{1}{2}}+6\right)^2}=\frac{6x+27\sqrt{x}+\frac{1}{\sqrt{x}}}{\left(2x^{\frac{1}{2}}+6\right)^2}=\frac{6x\sqrt{x}+27x+1}{\sqrt{x}(2)^2(\sqrt{x}+3)^2}=$$

$$\frac{6x\sqrt{x}+27x+1}{4\sqrt{x}(\sqrt{x}+3)^2}.$$

5. If $y=\dfrac{-15}{x}$, then $y'=\dfrac{x\dfrac{d}{dx}(-15)-(-15)\dfrac{d}{dx}(x)}{(x)^2}=\dfrac{x(0)-(-15)(1)}{(x)^2}=\dfrac{15}{x^2}$.

6. If $s(t)=\dfrac{2t^{\frac{3}{2}}-3}{4t^{\frac{1}{2}}+6}$, then $s'(t)=\dfrac{\left(4t^{\frac{1}{2}}+6\right)\dfrac{d}{dt}\left(2t^{\frac{3}{2}}-3\right)-\left(2t^{\frac{3}{2}}-3\right)\dfrac{d}{dt}\left(4t^{\frac{1}{2}}+6\right)}{\left(4x^{\frac{1}{2}}+6\right)^2}=\dfrac{\left(4t^{\frac{1}{2}}+6\right)\left(3t^{\frac{1}{2}}\right)-\left(2t^{\frac{3}{2}}-3\right)\left(2t^{-\frac{1}{2}}\right)}{\left(4x^{\frac{1}{2}}+6\right)^2}=$

$$\frac{(12t+18\sqrt{t})-\left(4t-\frac{6}{\sqrt{t}}\right)}{\left(4x^{\frac{1}{2}}+6\right)^2}=\frac{8t+18\sqrt{t}+\frac{6}{\sqrt{t}}}{\left(4x^{\frac{1}{2}}+6\right)^2}=\frac{8t\sqrt{t}+18t+6}{\sqrt{t}(2)^2(2\sqrt{t}+3)^2}=\frac{4t\sqrt{t}+9t+3}{2\sqrt{t}(2\sqrt{t}+3)^2}.$$

7. If $g(x)=\dfrac{x^{100}}{x^{-5}+10}$, then $g'(x)=\dfrac{(x^{-5}+10)\dfrac{d}{dx}(x^{100})-(x^{100})\dfrac{d}{dx}(x^{-5}+10)}{(x^{-5}+10)^2}=$

$$\frac{(x^{-5}+10)(100x^{99})-(x^{100})(-5x^{-6})}{(x^{-5}+10)^2}=\frac{105x^{94}+1000x^{99}}{\left(\frac{1}{x^5}+10\right)^2}.$$

8. If $y=\dfrac{4-5x^3}{8x^2-7}$, then $y'=\dfrac{(8x^2-7)\dfrac{d}{dx}(4-5x^3)-(4-5x^3)\dfrac{d}{dx}(8x^2-7)}{(8x^2-7)^2}=$

$$\frac{(8x^2-7)(-15x^2)-(4-5x^3)(16x)}{(8x^2-7)^2}=\frac{-40x^4+105x^2-64x}{(8x^2-7)^2}.$$

9. If $q(v)=\dfrac{v^3+2}{v^2-\dfrac{1}{v^3}}$, then $q'(v)=\dfrac{\left(v^2-\dfrac{1}{v^3}\right)\dfrac{d}{dv}(v^3+2)-(v^3+2)\dfrac{d}{dv}\left(v^2-\dfrac{1}{v^3}\right)}{\left(v^2-\dfrac{1}{v^3}\right)^2}=$

$$\frac{\left(v^2-\frac{1}{v^3}\right)(3v^2)-(v^3+2)(2v+3v^{-4})}{\left(v^2-\frac{1}{v^3}\right)^2}=\frac{v^4-\frac{6}{v}-4v-\frac{6}{v^4}}{\left(v^2-\frac{1}{v^3}\right)^2}=\frac{v^8-6v^3-4v^5-6}{v^4\left(v^2-\frac{1}{v^3}\right)^2}.$$

10. If $f(x)=\dfrac{-4x^2}{\dfrac{4}{x^2}+8}$, then $f'(x)=\dfrac{(4x^{-2}+8)\dfrac{d}{dx}(-4x^2)-(-4x^2)\dfrac{d}{dx}(4x^{-2}+8)}{(4x^{-2}+8)^2}=$

$$\frac{(4x^{-2}+8)(-8x)-(-4x^2)(-8x^{-3})}{(4x^{-2}+8)^2}=\frac{-32x^{-1}-64x-32x^{-1}}{(4x^{-2}+8)^2}=\frac{\frac{-64}{x}-64x}{(4x^{-2}+8)^2}=\frac{-64-64x^2}{x\left(\frac{4}{x^2}+8\right)^2}.$$

11. If $f(x)=\frac{5x+2}{3x-1}$, then $f'(x)=-\frac{11}{(3x-1)^2}$; thus, $f'(25)=-\frac{11}{(3(25)-1)^2}=-\frac{11}{5476}$.

12. If $h(x)=\frac{4-5x^2}{8x}$, then $h'(x)=-\frac{1}{2x^2}-\frac{5}{8}$; thus, $h'(0.2)=-\frac{1}{2(0.2)^2}-\frac{5}{8}=-13.125$.

13. If $g(x)=\frac{5}{\sqrt{x}}$, then $g'(x)=-\frac{5}{2x^{\frac{3}{2}}}$; thus, $g'(0.25)=-\frac{5}{2(0.25)^{\frac{3}{2}}}=-20$.

14. If $y=\frac{-15}{x}$, then $y'=\frac{15}{x^2}$; thus, $\left.\frac{dy}{dx}\right|_{10}=\frac{15}{(10)^2}=0.15$.

15. If $g(x)=\frac{x^{100}}{x^{-5}+10}$, then $g'(x)=\frac{105x^{94}+1000x^{99}}{\left(\frac{1}{x^5}+10\right)^2}$; thus, $g'(1)=\frac{105(1)^{94}+1000(1)^{99}}{\left(\frac{1}{(1)^5}+10\right)^2}=\frac{1105}{121}$.

**5·5**

1. $f'(x)=\frac{d}{dx}[(3x^2-10)^3]=3(3x^2-10)^2\frac{d}{dx}(3x^2-10)=3(3x^2-10)^2(6x)=18x(3x^2-10)^2$.

2. $g'(x)=\frac{d}{dx}[40(3x^2-10)^3]=40\frac{d}{dx}[(3x^2-10)^3]=40(18x(3x^2-10)^2)=720x(3x^2-10)^2$.

3. $h'(x)=\frac{d}{dx}[10(3x^2-10)^{-3}]=10(-3)(3x^2-10)^{-4}\frac{d}{dx}(3x^2-10)=\left(\frac{-30}{(3x^2-10)^4}\right)(6x)=-\frac{180x}{(3x^2-10)^4}$.

4. $h'(x)=\frac{d}{dx}[(\sqrt{x}+3)^2]=2(\sqrt{x}+3)\frac{d}{dx}[(\sqrt{x}+3)]=2(\sqrt{x}+3)\left(\frac{1}{2\sqrt{x}}\right)=1+\frac{3}{\sqrt{x}}$.

5. $f'(u)=\frac{d}{du}\left[\left(\frac{1}{u^2}-u\right)^3\right]=3\left(\frac{1}{u^2}-u\right)^2\frac{d}{du}(u^{-2}-u)=3\left(\frac{1}{u^2}-u\right)^2(-2u^{-3}-1)=-\left(\frac{6}{u^3}+3\right)\left(\frac{1}{u^2}-u\right)^2$.

6. $y'=\frac{d}{dx}\left[\frac{1}{(x^2-8)^3}\right]=\frac{d}{dx}[(x^2-8)^{-3}]=-3(x^2-8)^{-4}\frac{d}{dx}(x^2-8)=-3(x^2-8)^{-4}(2x)=\frac{-6x}{(x^2-8)^4}$.

7. $y'=\frac{d}{dx}\sqrt{2x^3+5x+1}=\frac{d}{dx}(2x^3+5x+1)^{\frac{1}{2}}=\frac{1}{2}(2x^3+5x+1)^{-\frac{1}{2}}\frac{d}{dx}(2x^3+5x+1)=\frac{1}{2}(2x^3+5x+1)^{-\frac{1}{2}}(6x^2+5)=$

$\frac{6x^2+5}{2(2x^3+5x+1)^{\frac{1}{2}}}=\frac{6x^2+5}{2\sqrt{2x^3+5x+1}}$.

8. $s'(t)=\frac{d}{dt}\left[(2t^3+5t)^{\frac{1}{3}}\right]=\frac{1}{3}(2t^3+5t)^{-\frac{2}{3}}\frac{d}{dt}(2t^3+5t)=\frac{1}{3}(2t^3+5t)^{-\frac{2}{3}}(6t^2+5)=\frac{6t^2+5}{3(2t^3+5t)^{\frac{2}{3}}}$.

9. $f'(x)=\frac{d}{dx}\left[\frac{10}{(2x-6)^5}\right]=10\frac{d}{dx}[(2x-6)^{-5}]=10(-5)(2x-6)^{-6}\frac{d}{dx}(2x-6)=(-50)(2x-6)^{-6}(2)=-\frac{100}{(2x-6)^6}$.

10. $C'(t)=\frac{d}{dt}\left(\frac{50}{\sqrt{15t+120}}\right)=50\frac{d}{dt}\left((15t+120)^{-\frac{1}{2}}\right)=50\left(-\frac{1}{2}\right)(15t+120)^{-\frac{3}{2}}\frac{d}{dt}(15t+120)=$

$(-25)(15t+120)^{-\frac{3}{2}}(15)=-\frac{375}{(15t+120)^{\frac{3}{2}}}$.

11. $f'(x)=\frac{d}{dx}[(3x^2-10)^3]=18x(3x^2-10)^2$; thus, $f'(10)=18(10)(3(10)^2-10)^2=1.5138\cdot 10^7$.

12. $h'(x)=\frac{d}{dx}[10(3x^2-10)^{-3}]=-\frac{180x}{(3x^2-10)^4}$; thus, $h'(3)=-\frac{180(3)}{(3(3)^2-10)^4}=-\frac{540}{83{,}521}$.

13. $f'(x)=\frac{d}{dx}[(\sqrt{x}+3)^2]=1+\frac{3}{\sqrt{x}}$; thus, $f'(144)=1+\frac{3}{\sqrt{144}}=1\frac{1}{4}$.

14. $f'(u)=\frac{d}{du}\left[\left(\frac{1}{u^2}-u\right)^3\right]=-\left(\frac{6}{u^3}+3\right)\left(\frac{1}{u^2}-u\right)^2$; thus, $f'(2)=-\left(\frac{6}{(2)^3}+3\right)\left(\frac{1}{(2)^2}-(2)\right)^2=-\frac{735}{64}$.

15. $y'=\frac{d}{dx}\left[\frac{1}{(x^2-8)^3}\right]=\frac{-6x}{(x^2-8)^4}$; thus, $\left.\frac{dy}{dx}\right|_4=\frac{-6(4)}{((4)^2-8)^4}=\frac{-3}{512}$.

**5·6**

1. Step 1. Differentiate every term on both sides of the equation with respect to $x$:

$$\frac{d}{dx}(x^2y)=\frac{d}{dx}(1)$$

$$x^2\frac{d}{dx}(y)+y\frac{d}{dx}(x^2)=0$$

$$x^2\frac{dy}{dx}+2xy=0$$

Step 2. Solve the resulting equation for $\frac{dy}{dx}$.

$$x^2\frac{dy}{dx}=-2xy$$

$$\frac{dy}{dx}=\frac{-2xy}{x^2}=\frac{-2y}{x}, \text{ provided } x\neq 0$$

2. Step 1. Differentiate every term on both sides of the equation with respect to $x$:

$$\frac{d}{dx}(xy^3)=\frac{d}{dx}(3x^2y+5y)$$

$$x\frac{d}{dx}(y^3)+y^3\frac{d}{dx}(x)=3x^2\frac{d}{dx}(y)+y\frac{d}{dx}(3x^2)+5\frac{d}{dx}(y)$$

$$x(3y^2)\frac{dy}{dx}+y^3(1)=3x^2\frac{dy}{dx}+y(6x)+5\frac{dy}{dx}$$

$$(3xy^2)\frac{dy}{dx}+y^3=3x^2\frac{dy}{dx}+6xy+5\frac{dy}{dx}$$

Step 2. Solve the resulting equation for $\frac{dy}{dx}$.

$$(3xy^2)\frac{dy}{dx}-3x^2\frac{dy}{dx}-5\frac{dy}{dx}=6xy-y^3$$

$$\frac{dy}{dx}=\frac{6xy-y^3}{3xy^2-3x^2-5}.$$

3. Step 1. Differentiate every term on both sides of the equation with respect to $x$:

$$\frac{d}{dx}\left(x^{\frac{1}{2}}\right)+\frac{d}{dx}\left(y^{\frac{1}{2}}\right)=\frac{d}{dx}(25)$$

$$\frac{1}{2}x^{-\frac{1}{2}}+\left(\frac{1}{2}y^{-\frac{1}{2}}\right)\frac{dy}{dx}=0$$

Step 2. Solve the resulting equation for $\frac{dy}{dx}$.

$$\frac{dy}{dx}=-\frac{\frac{1}{2}x^{-\frac{1}{2}}}{\frac{1}{2}y^{-\frac{1}{2}}}=-\frac{\sqrt{y}}{\sqrt{x}}.$$

4. Step 1. Differentiate every term on both sides of the equation with respect to $x$:

$$\frac{d}{dx}(x^{-1})+\frac{d}{dx}(y^{-1})=\frac{d}{dx}(9)$$

$$-1x^{-2}+(-1y^{-2})\frac{dy}{dx}=0$$

Step 2. Solve the resulting equation for $\frac{dy}{dx}$.

$$\frac{dy}{dx}=-\frac{x^{-2}}{y^{-2}}=-\frac{y^2}{x^2}$$

5. Step 1. Differentiate every term on both sides of the equation with respect to $x$:

$$\frac{d}{dx}(x^2)+\frac{d}{dx}(y^2)=\frac{d}{dx}(16)$$

$$2x+(2y)\frac{dy}{dx}=0$$

Step 2. Solve the resulting equation for $\frac{dy}{dx}$.

$$\frac{dy}{dx}=-\frac{2x}{2y}=-\frac{x}{y}$$

6. When $x^2y=1$, $\frac{dy}{dx}=\frac{-2xy}{x^2}=\frac{-2y}{x}$; thus, $\left.\frac{dy}{dx}\right|_{(3,1)}=\frac{-2(1)}{(3)}=-\frac{2}{3}$.

7. When $xy^3=3x^2y+5y$, $\frac{dy}{dx}=\frac{6xy-y^3}{3xy^2-3x^2-5}$; thus, $\left.\frac{dy}{dx}\right|_{(5,2)}=\frac{6(5)(2)-(2)^3}{3(5)(2)^2-3(5)^2-5}=-\frac{13}{5}$.

8. When $\sqrt{x}+\sqrt{y}=25$, $\frac{dy}{dx}=-\frac{\sqrt{y}}{\sqrt{x}}$; thus, $\left.\frac{dy}{dx}\right|_{(4,9)}=-\frac{\sqrt{9}}{\sqrt{4}}=-\frac{3}{2}$.

9. When $\frac{1}{x}+\frac{1}{y}=9$, $\frac{dy}{dx}=-\frac{y^2}{x^2}$; thus, $\left.\frac{dy}{dx}\right|_{(5,10)}=-\frac{100}{25}=-4$.

10. When $x^2+y^2=16$, $\frac{dy}{dx}=-\frac{x}{y}$; thus, $\left.\frac{dy}{dx}\right|_{(2,1)}=-\frac{(2)}{(1)}=-2$.

# 6 Additional derivatives

**6·1**

1. $f'(x)=\frac{d}{dx}(20e^x)=20\cdot\frac{d}{dx}(e^x)=20e^x$
2. $y'=\frac{d}{dx}(e^{3x})=e^{3x}(3)=3e^{3x}$
3. $g'(x)=\frac{d}{dx}=e^{5x^3}(15x^2)=15x^2e^{5x^3}$
4. $y'=\frac{d}{dx}(-4e^{5x^3})=-4\cdot(e^{5x^3})(15x^2)=-60x^2e^{5x^3}$
5. $h'(x)=\frac{d}{dx}(e^{-10x^3})=e^{-10x^3}(-30x^2)=-30x^2e^{-10x^3}$
6. $f'(x)=15\cdot\frac{d}{dx}(x^2)+10\cdot\frac{d}{dx}(e^x)=15\cdot 2x+10(e^x)=30x+10e^x$

7. $g'(x)=\frac{d}{dx}(e^{7x-2x^3})=e^{7x-2x^3}(7-6x^2)=(7-6x^2)e^{7x-2x^3}$

8. $f'(t)=\frac{d}{dt}\left(\frac{100}{e^{-0.5t}}\right)=100\frac{d}{dt}(e^{0.5t})=100e^{0.5t}(0.5)=50e^{0.5t}$

9. $g'(t)=2500\cdot\frac{d}{dt}(e^{2t+1})=2500e^{2t+1}\cdot(2)=5000e^{2t+1}$

10. $f'(x)=\frac{1}{\sqrt{2\pi}}\cdot\frac{d}{dx}\left(e^{-\frac{x^2}{2}}\right)=\frac{1}{\sqrt{2\pi}}\cdot e^{-\frac{x^2}{2}}(-x)=-\frac{x}{\sqrt{2\pi}}\cdot e^{-\frac{x^2}{2}}$

**6·2**

1. $f'(x)=20\cdot\frac{d}{dx}(\ln x)=20\cdot\frac{1}{x}=\frac{20}{x}$

2. $y'=\frac{d}{dx}(\ln 3x)=\frac{1}{3x}(3)=\frac{1}{x}$

3. $g'(x)=\frac{d}{dx}[\ln(5x^3)]=\frac{1}{5x^3}(15x^2)=\frac{3}{x}$

4. $y'=-4\cdot\frac{d}{dx}[\ln(5x^3)]=-4\cdot\frac{1}{5x^3}(15x^2)=-\frac{12}{x}$

5. $h'(x)=\frac{d}{dx}[\ln(-10x^3)]=-\frac{1}{10x^3}(-30x^2)=\frac{3}{x}$

6. $f'(x)=15\cdot\frac{d}{dx}(x^2)+10\cdot\frac{d}{dx}(\ln x)=15\cdot(2x)+10\cdot\frac{1}{x}=30x+\frac{10}{x}$

7. $g'(x)=\frac{d}{dx}[\ln(7x-2x^3)]=\frac{1}{7x-2x^3}(7-6x^2)=\frac{7-6x^2}{7x-2x^3}$

8. $f'(t)=\frac{d}{dt}[\ln(3t^2+5t-20)]=\frac{1}{3t^2+5t-20}(6t+5)=\frac{6t+5}{3t^2+5t-20}$

9. $g'(t)=\frac{d}{dt}[\ln(e^t)]=\frac{d}{dt}(t)=1$

10. $f'(x)=\frac{d}{dx}[\ln(\ln x)]=\frac{1}{\ln x}\cdot\frac{1}{x}=\frac{1}{x\ln x}$

**6·3**

1. $f'(x)=20\cdot\frac{d}{dx}(3^x)=20(\ln 3)(3^x)$

2. $y'=\frac{d}{dx}(5^{3x})=(\ln 5)(5^{3x})(3)=3(\ln 5)(5^{3x})$

3. $g'(x)=\frac{d}{dx}(2^{5x^3})=15x^2(\ln 2)(2^{5x^3})$

4. $y'=-4\cdot\frac{d}{dx}(2^{5x^3})=-60x^2(\ln 2)(2^{5x^3})$

5. $h'(x)=\frac{d}{dx}(4^{-10x^3})=-30x^2(\ln 4)(4^{-10x^3})$

6. $f'(x)=15\cdot\frac{d}{dx}(x^2)+10\cdot\frac{d}{dx}(5^{3x})=15\cdot(2x)+10(\ln 5)(5^{3x})\cdot 3=30x+30(\ln 5)(5^{3x})$

7. $g'(x)=\frac{d}{dx}(3^{7x-2x^3})=(\ln 3)(3^{7x-2x^3})(7-6x^2)$

8. $f'(t)=100\cdot\frac{d}{dt}(10^{0.5t})=100(\ln 10)(10^{0.5t})(0.5)=50(\ln 10)(10^{0.5t})$

9. $g'(t) = 2500 \cdot \frac{d}{dt}(5^{2t+1}) = 2500(\ln 5)(5^{2t+1})(2) = 5000(\ln 5)(5^{2t+1})$

10. $f'(x) = \frac{d}{dx}\left(8^{-\frac{x^2}{2}}\right) = (\ln 8)\left(8^{-\frac{x^2}{2}}\right)\left(-\frac{1}{2} \cdot 2x\right) = -x(\ln 8)\left(8^{-\frac{x^2}{2}}\right)$

**6·4**

1. $f'(x) = 20 \cdot \frac{d}{dx}(\log_4 x) = 20 \cdot \frac{1}{x \ln 4} = \frac{20}{x \ln 4}$

2. $y' = \frac{d}{dx}(\log_{10} 3x) = \frac{1}{3x \ln 10}(3) = \frac{1}{x \ln 10}$

3. $g'(x) = \frac{d}{dx}[\log_8(5x^3)] = \frac{1}{5x^3 \ln 8}(15x^2) = \frac{3}{x \ln 8}$

4. $y' = \frac{d}{dx}[-4\log_8(5x^3)] = -4 \cdot \frac{d}{dx}[\log_8(5x^3)] = -\frac{12}{x \ln 8}$

5. $h'(x) = \frac{d}{dx}[\log_5(-10x^3)] = \frac{1}{-10x^3 \ln 5}(-30x^2) = \frac{3}{x \ln 5}$

6. $f'(x) = 15 \cdot \frac{d}{dx}(x^2) + 10 \cdot \frac{d}{dx}(\log_2 x) = 30x + \frac{10}{x \ln 2}$

7. $g'(x) = \frac{d}{dx}[\log_6(7x - 2x^3)] = \frac{1}{(7x - 2x^3)\ln 6}(7 - 6x^2) = \frac{7 - 6x^2}{(7x - 2x^3)\ln 6}$

8. $f'(t) = \frac{d}{dt}[\log_{16}(3t^2 + 5t - 20)] = \frac{1}{(3t^2 + 5t - 20)\ln 16}(6t + 5) = \frac{6t + 5}{(3t^2 + 5t - 20)\ln 16}$

9. $g'(t) = \frac{d}{dt}[\log_2(e^t)] = \frac{1}{e^t \ln 2}(e^t) = \frac{1}{\ln 2}$

10. $f'(x) = \frac{d}{dx}[\log_{10}(\log_{10} x)] = \frac{1}{(\ln 10)(\log_{10} x)} \frac{d}{dx}(\log_{10} x) = \frac{1}{(\ln 10)\left(\frac{\ln x}{\ln 10}\right)} \frac{d}{dx}(\log_{10} x) = \frac{1}{\ln x} \cdot \frac{1}{x \ln 10} =$

$\frac{1}{x(\ln 10)(\ln x)}$

**6·5**

1. $f'(x) = 5 \cdot \frac{d}{dx}(\sin 3x) = 5 \cdot (\cos 3x)(3) = 15\cos 3x$

2. $h'(x) = \frac{1}{4} \cdot \frac{d}{dx}[\cos(2x^2)] = -\frac{1}{4}\sin(2x^2) \cdot (4x) = -x\sin(2x^2)$

3. $g'(x) = 5 \cdot \frac{d}{dx}\left[\tan\left(\frac{3x}{5}\right)\right] = 5\sec^2\left(\frac{3x}{5}\right) \cdot \left(\frac{3}{5}\right) = 3\sec^2\left(\frac{3x}{5}\right)$

4. $f'(x) = 10 \cdot \frac{d}{dx}(\sec 2x) = 10\sec 2x \tan 2x \cdot (2) = 20\sec 2x \tan 2x$

5. $y' = \frac{2}{3} \cdot \frac{d}{dx}[\sec(2x^3)] = \frac{2}{3}\sec(2x^3)\tan(2x^3) \cdot (6x^2) = 4x^2\sec(2x^3)\tan(2x^3)$

6. $s'(t) = 4 \cdot \frac{d}{dt}(\cot 5t) = -4\csc^2 5t \cdot (5) = -20\csc^2 5t$

7. $g'(x) = 6 \cdot \frac{d}{dx}\left[\tan^3\left(\frac{2x}{3}\right)\right] - 20 \cdot \frac{d}{dx}\left(x^{\frac{1}{2}}\right) = 6 \cdot 3\tan^2\left(\frac{2x}{3}\right) \cdot \sec^2\left(\frac{2x}{3}\right) \cdot \left(\frac{2}{3}\right) -$

$20 \cdot \frac{1}{2}\left(x^{-\frac{1}{2}}\right) = 12\tan^2\left(\frac{2x}{3}\right)\sec^2\left(\frac{2x}{3}\right) - \frac{10}{\sqrt{x}}$

8. $f'(x)=2\cdot\frac{d}{dx}(x\sin x)+\frac{d}{dx}(\cos 2x)=2[x(\cos x)+\sin x(1)]+-\sin 2x(2)=2x\cos x+2\sin x-2\sin 2x$

9. $h'(x)=\frac{d}{dx}\left[\frac{\sin 3x}{1+\sin 3x}\right]=\frac{(1+\sin 3x)(\cos 3x)(3)-\sin 3x(0+\cos 3x)(3)}{(1+\sin 3x)^2}=$

$\frac{3(1+\sin 3x)(\cos 3x)-3\sin 3x\cos 3x}{(1+\sin 3x)^2}=\frac{3\cos 3x+3\sin 3x\cos 3x-3\sin 3x\cos 3x}{(1+\sin 3x)^2}=\frac{3\cos 3x}{(1+\sin 3x)^2}$

10. $f'(x)=\frac{d}{dx}[e^{4x}\sin 2x]=e^{4x}(\cos 2x)(2)+(\sin 2x)(e^{4x})(4)=2e^{4x}\cos 2x+4e^{4x}\sin 2x$

**6·6**

1. $f'(x)=\frac{d}{dx}[\sin^{-1}(-x^3)]=\frac{1}{\sqrt{1-(-x^3)^2}}\cdot(-3x^2)=-\frac{3x^2}{\sqrt{1-x^6}}$

2. $h'(x)=\frac{d}{dx}[\cos^{-1}(e^x)]=-\frac{1}{\sqrt{1-(e^x)^2}}\cdot(e^x)=-\frac{e^x}{\sqrt{1-e^{2x}}}$

3. $g'(x)=\frac{d}{dx}[\tan^{-1}(x^2)]=\frac{1}{1+(x^2)^2}\cdot(2x)=\frac{2x}{1+x^4}$

4. $f'(x)=\frac{d}{dx}[\cot^{-1}(7x-5)]=-\frac{1}{1+(7x-5)^2}\cdot(7)=-\frac{7}{1+(7x-5)^2}$

5. $y'=\frac{1}{15}\cdot\frac{d}{dx}[\sin^{-1}(5x^3)]=\frac{1}{15}\cdot\frac{1}{\sqrt{1-(5x^3)^2}}\cdot(15x^2)=\frac{x^2}{\sqrt{1-25x^6}}$

6. $f'(x)=\frac{d}{dx}[\cos^{-1}(x^2)]=-\frac{1}{\sqrt{1-(x^2)^2}}\cdot(2x)=-\frac{2x}{\sqrt{1-x^4}}$

7. $h'(x)=\frac{d}{dx}[\csc^{-1}(2x)]=-\frac{1}{|2x|\sqrt{(2x)^2-1}}\cdot(2)=-\frac{1}{|x|\sqrt{4x^2-1}}$

8. $g'(x)=4\cdot\frac{d}{dx}\left[\sec^{-1}\left(\frac{x}{2}\right)\right]=4\cdot\frac{1}{\left|\frac{x}{2}\right|\sqrt{\left(\frac{x}{2}\right)^2-1}}\cdot\left(\frac{1}{2}\right)=\frac{4}{|x|\sqrt{\frac{x^2}{4}-1}}$

9. $f'(x)=\frac{d}{dx}[x\sin^{-1}(7x^2)]=\left[x\frac{1}{\sqrt{1-(7x^2)^2}}(14x)+\sin^{-1}(7x^2)(1)\right]=\frac{14x^2}{\sqrt{1-49x^4}}+\sin^{-1}(7x^2)$

10. $y'=\frac{d}{dx}[\arcsin(\sqrt{1-x^2})]=\frac{1}{\sqrt{1-(\sqrt{1-x^2})^2}}\cdot\frac{1}{2}(1-x^2)^{-\frac{1}{2}}(-2x)=-\frac{x}{\sqrt{1-x^2}\sqrt{1-(1-x^2)}}=$

$-\frac{x}{\sqrt{1-x^2}\sqrt{1-1+x^2}}=-\frac{x}{\sqrt{1-x^2}\sqrt{x^2}}=-\frac{x}{|x|\sqrt{1-x^2}}$

**6·7**

1. $f'(x)=\frac{d}{dx}[x^7+2x^{10}]=7x^6+20x^9; f''(x)=42x^5+180x^8;$ thus, $f'''(x)=210x^4+1440x^7$

2. $h'(x)=\frac{d}{dx}x^{\frac{1}{3}}=\frac{1}{3}x^{\frac{-2}{3}};$ thus, $h''(x)=-\frac{2}{9}x^{-\frac{5}{3}}=-\frac{2}{9x^{\frac{5}{3}}}$

3. $g'(x)=\frac{d}{dx}(2x)=2; g''(x)=0;$ thus, $g^{(5)}(x)=0$

4. $f'(x)=5\cdot\frac{d}{dx}(e^x)=5e^x; f''(x)=5e^x; f'''(x)=5e^x; f^{(4)}(x)=5e^x$

5. $y'=\frac{d}{dx}(\sin 3x)=3\cos 3x; \frac{d^2y}{d^2x}=-9\sin 3x;$ thus, $\frac{d^3y}{d^3x}=-27\cos 3x$

6. $s'(t)=\frac{d}{dt}\left[16t^2-\frac{2t}{3}+10\right]=32t-\frac{2}{3};$ thus, $s''(t)=32$

7. $g'(x)=D_x[\ln 3x]=\frac{3}{3x}=\frac{1}{x}; D_x^2[g(x)]=-\frac{1}{x^2};$ thus, $D_x^3[g(x)]=\frac{2}{x^3}$

8. $f'(x)=10\cdot\frac{d}{dx}(x^{-5})+\frac{1}{5}\cdot\frac{d}{dx}(x^3); f'(x)=-50x^{-6}+\frac{3}{5}x^2; f''(x)=300x^{-7}+\frac{6}{5}x; f'''(x)=-2100x^{-8}+\frac{6}{5};$

thus, $f^{(4)}(x)=16{,}800x^{-9}+0=\frac{16{,}800}{x^9}$

9. $f'(x)=\frac{d}{dx}(3^{2x})=2\ln 3(3^{2x}); f''(x)=4(\ln 3)^2(3^{2x});$ thus, $f'''(x)=8(\ln 3)^3(3^{2x})$

10. $y'=\frac{dy}{dx}=\frac{d}{dx}(\log_2 5x)=\frac{1}{5x\ln 2}(5)=\frac{1}{x\ln 2}; \frac{d^2y}{d^2x}=\frac{1}{\ln 2}\frac{d}{dx}\left(\frac{1}{x}\right)=-\frac{1}{x^2\ln 2};$

$\frac{d^3y}{d^3x}=\frac{1}{\ln 2}\frac{d}{dx}\left(-\frac{1}{x^2}\right)=\frac{2}{x^3\ln 2};$ thus, $\frac{d^4y}{d^4x}=\frac{1}{\ln 2}\frac{d}{dx}\left(\frac{2}{x^3}\right)=-\frac{6}{x^4\ln 2}$

# INTEGRATION

## 7 Indefinite integral and basic integration formulas and rules

**7·1**

1. $\frac{d}{dx}(100x+C)=100+0=100$

2. $\frac{d}{dx}(3x^2+C)=6x+0=6x$

3. $\frac{d}{dx}(x^3+2x^2-5x+C)=3x^2+4x-5+0=3x^2+4x-5$

4. $\frac{d}{dx}\left(\frac{2}{7}x^{\frac{7}{2}}+\frac{2}{3}x^{\frac{3}{2}}+C\right)=\frac{7}{2}\cdot\frac{2}{7}x^{\frac{5}{2}}+\frac{3}{2}\cdot\frac{2}{3}x^{\frac{1}{2}}+0=x^{\frac{5}{2}}+x^{\frac{1}{2}}=\left(x^{\frac{4}{2}}+1\right)x^{\frac{1}{2}}=(x^2+1)\sqrt{x}$

5. $\frac{d}{dx}\left(\frac{x^{e+1}}{e+1}+e^x+C\right)=\frac{1}{e+1}(e+1)x^e+e^x+0=x^e+e^x$

6. $\frac{d}{dx}\left[\frac{(10x+30)^4}{4}+C\right]=4\cdot\frac{1}{4}\cdot(10x+30)^3(10)+0=10(10x+30)^3$

7. $\frac{d}{dx}\left[\frac{(x^2-3)^5}{5}+C\right]=5\cdot\frac{1}{5}\cdot(x^2-3)^4(2x)=(x^2-3)^4(2x)$

8. $\frac{d}{dx}\left(\frac{\sin^3 x}{3}+C\right)=3\cdot\frac{1}{3}\cdot\sin^2 x(\cos x)+0=\sin^2 x(\cos x)$

9. $\frac{d}{dx}\left(-\frac{\cos x^3}{3}+C\right)=\left(-\frac{1}{3}\right)(-\sin x^3)(3x^2)+0=x^2(\sin x^3)$

10. $\frac{d}{dx}(x\ln x-x+C)=x\cdot\frac{1}{x}+\ln x\cdot 1-1+0=1+\ln x-1=\ln x$

**7·2**

1. $\int 8\,dx = 8x + C$
2. $\int \frac{3}{4}dx = \frac{3}{4}x + C$
3. $\int 9.75\,dx = 9.75x + C$
4. $\int \sqrt{3}\,dx = \sqrt{3}x + C$
5. $\int \left(\frac{\sqrt[3]{40}}{\sqrt{10+15}}\right)dx = \frac{\sqrt[3]{40}}{\sqrt{10+15}}x + C$
6. $\int 16\sqrt{2}\,dt = 16\sqrt{2}t + C$
7. $\int e^2 dx = e^2 x + C$
8. $\int 2\pi\,dr = 2\pi r + C$
9. $\int -21\,du = -21u + C$
10. $\int \frac{6}{e}\,dx = \frac{6}{e}x + C$

**7·3**

1. $\int x^5 dx = \frac{x^6}{6} + C$
2. $\int \sqrt[4]{x^3}\,dx = \int x^{\frac{3}{4}}dx = \frac{x^{\frac{7}{4}}}{\frac{7}{4}} + C = \frac{4}{7}x^{\frac{7}{4}} + C$
3. $\int x^{\sqrt{2}}\,dx = \frac{x^{\sqrt{2}+1}}{\sqrt{2}+1} + C$
4. $\int \frac{1}{x^2}dx = \int x^{-2}dx = \frac{x^{-1}}{-1} + C = -\frac{1}{x} + C$
5. $\int t^{100}dt = \frac{t^{101}}{101} + C$
6. $\int u^{2\pi}\,du = \frac{u^{2\pi+1}}{2\pi+1} + C$
7. $\int \frac{1}{\sqrt{x}}\,dx = \int x^{-\frac{1}{2}}\,dx = \frac{x^{\frac{1}{2}}}{\frac{1}{2}} + C = 2x^{\frac{1}{2}} + C = 2\sqrt{x} + C$
8. $\int \frac{x^5}{x^2}dx = \int x^3 dx = \frac{x^4}{4} + C$
9. $\int r^{-1}dr = \ln|r| + C$
10. $\int \frac{1}{t}dt = \ln|t| + C$

**7·4**

1. $\int e^t dt = e^t + C$
2. $\int e^{20x}dx = \frac{e^{20x}}{20} + C$
3. $\int e^{\pi x}dx = \frac{e^{\pi x}}{\pi} + C$
4. $\int e^{0.25x}dx = \frac{e^{0.25x}}{0.25} + C = 4e^{0.25x} + C$

5. $\int e^{\frac{x}{5}}dx = \frac{e^{\frac{x}{5}}}{\frac{1}{5}} + C = 5e^{\frac{x}{5}} + C$

6. $\int e^{\sqrt{3}x}dx = \frac{e^{\sqrt{3}x}}{\sqrt{3}} + C$

7. $\int 4^x dx = \frac{4^x}{\ln 4} + C$

8. $\int 2^{3x} dx = \frac{2^{3x}}{3\ln 2} + C$

9. $\int 100^{0.25x} dx = \frac{100^{0.25x}}{0.25\ln 100} + C = \frac{4(100^{0.25x})}{\ln 100} + C$

10. $\int \pi^{\frac{x}{5}} dx = \frac{\pi^{\frac{x}{5}}}{\frac{1}{5}\ln \pi} + C = \frac{5\pi^{\frac{x}{5}}}{\ln \pi} + C$

**7·5**

1. $\int \cos v\, dv = \sin v + C$

2. $\int \sin\left(\frac{1}{2}\pi x\right)dx = \frac{-\cos\left(\frac{1}{2}\pi x\right)}{\frac{1}{2}\pi} + C = \frac{-2\cos\left(\frac{1}{2}\pi x\right)}{\pi} + C$

3. $\int \cos(18x)dx = \frac{\sin(18x)}{18} + C$

4. $\int \sec^2(\sqrt{3}x)dx = \frac{\tan(\sqrt{3}x)}{\sqrt{3}} + C$

5. $\int \csc^2(2.5x)dx = -\frac{\cot(2.5x)}{2.5} + C = -0.4\cot(2.5x) + C$

6. $\int \sec\left(\frac{5}{6}x\right)\tan\left(\frac{5}{6}x\right)dx = \frac{\sec\left(\frac{5}{6}x\right)}{\frac{5}{6}} + C = \frac{6}{5}\sec\left(\frac{5}{6}x\right) + C$

7. $\int \csc\frac{x}{3}\cot\frac{x}{3}dx = -\frac{\csc\left(\frac{x}{3}\right)}{\frac{1}{3}} + C = -3\csc\left(\frac{x}{3}\right) + C$

8. $\int \csc(ex)\cot(ex)dx = -\frac{\csc(ex)}{e} + C$

9. $\int \sin 3\theta\, d\theta = -\frac{\cos 3\theta}{3} + C$

10. $\int \cos(25\pi x)dx = \frac{\sin(25\pi x)}{25\pi} + C$

**7·6**

1. $\int \frac{1}{1+\theta^2}d\theta = \tan^{-1}\theta + C$

2. $\int \frac{dx}{\sqrt{16-x^2}} = \int \frac{1}{\sqrt{4^2-x^2}}dx = \sin^{-1}\left(\frac{x}{4}\right) + C$

3. $\int \frac{1}{49+x^2}dx = \int \frac{1}{7^2+x^2}dx = \frac{1}{7}\tan^{-1}\left(\frac{x}{7}\right) + C$

4. $\int \frac{dt}{0.25+t^2} = \int \frac{1}{(0.5)^2+t^2}dt = \frac{1}{0.5}\tan^{-1}\left(\frac{t}{0.5}\right) + C = 2\tan^{-1}(2t) + C$

5. $\int \frac{du}{\sqrt{u^2(u^2-1)}} = \int \frac{1}{|u|\sqrt{(u^2-1)}}du = \sec^{-1}u + C$

6. $\int \frac{1}{|x|\sqrt{x^2-41}}dx = \int \frac{1}{|x|\sqrt{x^2-(\sqrt{41})^2}}dx = \frac{1}{\sqrt{41}}\sec^{-1}\left(\frac{x}{\sqrt{41}}\right)+C$

7. $\int \frac{1}{\sqrt{\frac{81}{100}-x^2}}dx = \int \frac{1}{\sqrt{\left(\frac{9}{10}\right)^2-x^2}}dx = \sin^{-1}\left(\frac{x}{\frac{9}{10}}\right)+C = \sin^{-1}\left(\frac{10x}{9}\right)=C$

8. $\int \frac{1}{\pi^2+x^2}dx = \frac{1}{\pi}\tan^{-1}\left(\frac{x}{\pi}\right)+C$

9. $\int \frac{dt}{\sqrt{t^2\left(t^2-\frac{1}{4}\right)}} = \int \frac{1}{|t|\sqrt{\left(t^2-\left(\frac{1}{2}\right)^2\right)}}dt = \frac{1}{\frac{1}{2}}\sec^{-1}\left(\frac{t}{\frac{1}{2}}\right)+C = 2\sec^{-1}(2t)+C$

10. $\int \frac{1}{|x|\sqrt{x^2-7}}dx = \int \frac{1}{|x|\sqrt{x^2-(\sqrt{7})^2}}dx = \frac{1}{\sqrt{7}}\sec^{-1}\left(\frac{x}{\sqrt{7}}\right)+C$

**7·7**

1. $\int(3x^4-5x^3-21x^2+36x-10)dx = 3\int x^4dx - 5\int x^3dx - 21\int x^2dx + 36\int x\,dx - 10\int dx =$

$3\left(\frac{x^5}{5}\right)-5\left(\frac{x^4}{4}\right)-21\left(\frac{x^3}{3}\right)+36\left(\frac{x^2}{2}\right)-10(x)+C = \frac{3x^5}{5}-\frac{5x^4}{4}-7x^3+18x^2-10x+C$

2. $\int[3x^2-4\cos(2x)]\,dx = 3\int x^2dx - 4\int\cos(2x)dx = 3\cdot\frac{x^3}{3}-4\cdot\frac{\sin(2x)}{2}+C = x^3-2\sin(2x)+C$

3. $\int\left(\frac{8}{t^5}+\frac{5}{t}\right)dt = \int\frac{8}{t^5}dt+\int\frac{5}{t}dt = 8\int t^{-5}dt+5\int\frac{1}{t}dt = 8\left(\frac{t^{-4}}{-4}\right)+5\ln t+C = -\frac{2}{t^4}+5\ln t+C$

4. $\int\left(\frac{1}{\sqrt{25-\theta^2}}+\frac{1}{100+\theta^2}\right)d\theta = \int\frac{1}{\sqrt{5^2-\theta^2}}d\theta+\int\frac{1}{10^2+\theta^2}d\theta = \sin^{-1}\left(\frac{\theta}{5}\right)+\frac{1}{10}\tan^{-1}\left(\frac{\theta}{10}\right)+C$

5. $\int\frac{e^{5x}-e^{4x}}{e^{2x}}dx = \int(e^{3x}-e^{2x})dx = \int e^{3x}dx-\int e^{2x}dx = \frac{e^{3x}}{3}-\frac{e^{2x}}{2}+C$

6. $\int\left(\frac{x^7+x^4}{x^5}\right)dx = \int\left(x^2+\frac{1}{x}\right)dx = \int x^2dx+\int\frac{1}{x}dx = \frac{x^3}{3}+\ln x+C$

7. $\int\frac{1}{e^6+x^2}dx = \int\frac{1}{(e^3)^2+x^2}dx = \frac{1}{e^3}\tan^{-1}\left(\frac{x}{e^3}\right)+C$

8. $\int(x^2+4)^2dx = \int(x^4+8x^2+16)dx = \int x^4dx+8\int x^2dx+16\int dx = \frac{x^5}{5}+8\cdot\frac{x^3}{3}+16x+C = \frac{x^5}{5}+\frac{8x^3}{3}+16x+C$

9. $\int\left(\frac{7}{\sqrt[3]{t}}\right)dt = 7\int\left(t^{-\frac{1}{3}}\right)dt = 7\cdot\left(\frac{t^{\frac{2}{3}}}{\frac{2}{3}}\right)+C = \frac{21t^{\frac{2}{3}}}{2}+C$

10. $\int\frac{20+x}{\sqrt{x}}dx = \int\left(\frac{20}{\sqrt{x}}+\frac{x}{\sqrt{x}}\right)dx = 20\int x^{-\frac{1}{2}}dx+\int x^{\frac{1}{2}}dx+C = 20\cdot\left(\frac{x^{\frac{1}{2}}}{\frac{1}{2}}\right)+\frac{x^{\frac{3}{2}}}{\frac{3}{2}}+C =$

$40x^{\frac{1}{2}}+\frac{2}{3}x^{\frac{3}{2}}+C = 40\sqrt{x}+\frac{2}{3}|x|\sqrt{x}+C$

# 8 Basic integration techniques

**8·1**

1. $\int 3(x^3-5)^4 x^2 dx = \int (x^3-5)^4 3x^2 dx = \frac{(x^3-5)^5}{5} + C$

2. $\int e^{x^4} x^3 dx = \frac{1}{4}\int e^{x^4} 4x^3 dx = \frac{1}{4}e^{x^4} + C$

3. $\int \frac{t}{t^2+7} dt = \frac{1}{2}\int \frac{2t}{t^2+7} dt = \frac{1}{2}\ln(t^2+7) + C$

4. $\int (x^5-3x)^{\frac{1}{4}}(5x^4-3)dx = \frac{(x^5-3x)^{\frac{5}{4}}}{\frac{5}{4}} + C = \frac{4(x^5-3x)^{\frac{5}{4}}}{5} + C$

5. $\int \frac{x^3-2x}{(x^4-4x^2+5)^4} dx = \frac{1}{4}\int (x^4-4x^2+5)^{-4}(4x^3-8x)dx = \frac{1}{4}\cdot\frac{(x^4-4x^2+5)^{-3}}{-3} + C = -\frac{1}{12(x^4-4x^2+5)^3} + C$

6. $\int \frac{x^3-2x}{x^4-4x^2+5} dx = \frac{1}{4}\int \frac{1}{(x^4-4x^2+5)}(4x^3-8x)dx = \frac{1}{4}\ln|(x^4-4x^2+5)| + C$

7. $\int x\cos(3x^2+1)dx = \frac{1}{6}\int 6x\cos(3x^2+1)dx = \frac{1}{6}\sin(3x^2+1) + C$

8. $\int \frac{3\cos^2\sqrt{x}(\sin\sqrt{x})}{\sqrt{x}} dx = -2\int 3\cos^2\left(x^{\frac{1}{2}}\right)\left(-\sin x^{\frac{1}{2}}\right)\left(\frac{1}{2}x^{-\frac{1}{2}}\right)dx = -2\cos^3\left(x^{\frac{1}{2}}\right) + C = -2\cos^3(\sqrt{x}) + C$

9. $\int \frac{e^{2x}}{1+e^{4x}} dx = \frac{1}{2}\int \frac{1}{1+(e^{2x})^2} e^{2x} 2dx = \frac{1}{2}\tan^{-1}(e^{2x}) + C$

10. $\int 6t^2 e^{t^3-2} dt = 2\int e^{t^3-2} 3t^2 dt = 2e^{t^3-2} + C$

**8·2**

1. Let $u = x$ and $dv = \sin(2x)\cdot 2dx$, then $\int 2x\sin(2x)dx = \int x\sin(2x)2dx = u\cdot v - \int v\cdot du =$
$x(-\cos(2x)) - \int -\cos 2x\cdot dx = -x\cos(2x) + \frac{1}{2}\int \cos 2x\cdot 2\,dx = -x\cos(2x) + \frac{1}{2}\sin(2x) + C =$
$\frac{1}{2}\sin(2x) - x\cos(2x) + C.$

2. Let $u = \ln x$ and $dv = x^3 dx$, then $\int x^3 \ln x\,dx = u\cdot v - \int v\cdot du = \ln x\cdot\frac{x^4}{4} - \int \frac{x^4}{4}\cdot\frac{1}{x}dx =$
$\frac{x^4\ln x}{4} - \frac{1}{4}\int x^3 dx = \frac{x^4\ln x}{4} - \frac{1}{4}\cdot\frac{x^4}{4} + C = \frac{x^4\ln x}{4} - \frac{x^4}{16} + C.$

3. Let $u = t$ and $dv = e^t dt$, then $\int te^t dt = t\cdot e^t - \int e^t\cdot dt = te^t - e^t + C = e^t(t-1) + C.$

4. Let $u = x$ and $dv = \cos x\,dx$, then $\int x\cos x\,dx = x\cdot\sin x - \int \sin x\cdot dx = x\sin x + \cos x + C.$

5. Let $u = \cot^{-1} x$ and $dv = dx$, then $\int \cot^{-1}(x)dx = u\cdot v - \int v\cdot du = (\cot^{-1} x)\cdot x - \int x\cdot -\frac{1}{1+x^2}dx =$
$x(\cot^{-1} x) + \frac{1}{2}\int \frac{1}{(1+x^2)}2x\,dx = x(\cot^{-1} x) + \frac{1}{2}\ln|1+x^2| + C.$

6. Let $u = x^2$ and $dv = e^x dx$, then $\int x^2 e^x dx = u\cdot v - \int v\cdot du = x^2\cdot e^x - \int e^x\cdot 2x\,dx = x^2e^x - 2\int e^x x\,dx$; for the second integral, let $u = x$ and $dv = e^x dx$, then $\int x^2e^x dx = x^2e^x - 2\left(x\cdot e^x - \int e^x\cdot dx\right) = x^2e^x - 2xe^x + 2e^x + C.$

7. Let $u = w$ and $dv = (w-3)^2 dw$, then $\int w(w-3)^2 dw = u \cdot v - \int v \cdot du = w \cdot \frac{(w-3)^3}{3} - \int \frac{(w-3)^3}{3} \cdot dw =$ $\frac{w(w-3)^3}{3} - \frac{(w-3)^4}{12} + C$

8. Let $u = \ln(4x)$ and $dv = x^3 dx$, then $\int x^3 \ln(4x) dx = u \cdot v - \int v \cdot du = \ln(4x) \cdot \frac{x^4}{4} - \int \frac{x^4}{4} \cdot \frac{1}{x} dx =$ $\frac{x^4 \ln(4x)}{4} - \int \frac{x^3}{4} dx = \frac{x^4 \ln(4x)}{4} - \frac{x^4}{16} + C.$

9. Let $u = t$ and $dv = (t+5)^{-4} dt$, then $\int t(t+5)^{-4} dt = u \cdot v - \int v \cdot du = t \cdot \frac{(t+5)^{-3}}{-3} - \int -\frac{1}{3}(t+5)^{-3} \cdot dt =$ $-\frac{t}{3(t+5)^3} + \frac{1}{3}\int (t+5)^{-3} \cdot dt = -\frac{t}{3(t+5)^3} + \frac{1}{3} \cdot \frac{(t+5)^{-2}}{-2} = -\frac{t}{3(t+5)^3} - \frac{1}{6(t+5)^2} + C.$

10. Let $u = x$ and $dv = (x+2)^{\frac{1}{2}}$, then $\int x\sqrt{x+2}\, dx = x \cdot \left( \frac{2(x+2)^{\frac{3}{2}}}{3} \right) - \int \frac{2}{3}(x+2)^{\frac{3}{2}} \cdot dx = \frac{2x(x+2)^{\frac{3}{2}}}{3} -$ $\frac{2}{3}\int (x+2)^{\frac{3}{2}} dx = \frac{2x(x+2)^{\frac{3}{2}}}{3} - \frac{2}{3} \cdot \frac{2}{5}(x+2)^{\frac{5}{2}} + C = \frac{2x(x+2)^{\frac{3}{2}}}{3} - \frac{4}{15}(x+2)^{\frac{5}{2}} + C.$

**8·3**

1. $\int \cot x\, dx$ matches Formula 15. Therefore, $\int \cot x\, dx = \ln|\sin x| + C.$

2. $\int \frac{1}{(x+2)(3x+5)} dx$ matches Formula 50 with $a = 1$, $b = 2$, $c = 3$, and $d = 5$. Therefore, $\int \frac{1}{(x+2)(3x+5)} dx = \frac{1}{1 \cdot 5 - 2 \cdot 3} \ln\left|\frac{x+2}{3x+5}\right| + C = -\ln\left|\frac{x+2}{3x+5}\right| + C.$

3. $\int (\ln x)^2 dx$ matches Formula 41. Therefore, $\int (\ln x)^2 dx = 2x - 2x\ln x + x(\ln x)^2 + C.$

4. $\int x\cos x\, dx$ matches Formula 57. Therefore, $\int x\cos x\, dx = \cos x + x\sin x + C.$

5. $\int \frac{x}{(x+2)^2} dx$ matches Formula 49 with $a = 1$ and $b = 2$. Therefore, $\int \frac{x}{(x+2)^2} dx = \frac{1}{1^2}\left( \frac{2}{x+2} + \ln|x+2| \right)$ $C = \frac{2}{x+2} + \ln|x+2| + C.$

6. $\int 3xe^x dx = 3\int xe^x dx$ matches Formula 35. Therefore, $3\int xe^x dx = 3(e^x(x-1)) + C = 3e^x(x-1) + C.$

7. $\int \sqrt{10w+3}\, dw$ matches Formula 51 with $a = 10$ and $b = 3$. Therefore, $\int \sqrt{10w+3}\, dw = \frac{2}{3 \cdot 10}(10w+3)^{\frac{3}{2}} +$ $C = \frac{1}{15}(10w+3)^{\frac{3}{2}} + C.$

8. $\int t(t+5)^{-1} dt = \int \frac{t}{t+5} dt$ matches Formula 48 with $a = 1$ and $b = 5$. Therefore, $\int \frac{t}{t+5} dt = \frac{t}{1} - \frac{5}{1^2}\ln|t+5| +$ $C = t - 5\ln|t+5| + C.$

9. $\int x\sqrt{x+2}\, dx$ matches Formula 52 with $a = 1$ and $b = 2$. Therefore, $\int x\sqrt{x+2}\, dx = \frac{2(3 \cdot 1x - 2 \cdot 2)}{15 \cdot 1^2}(x+2)^{\frac{3}{2}} +$ $C = \frac{2(3x-4)}{15}(x+2)^{\frac{3}{2}} + C.$

10. $\int \frac{1}{\sin u \cos u} du$ matches Formula 58. Therefore, $\int \frac{1}{\sin u \cos u} du = \ln|\tan u| + C.$

# 9 The definite integral

**9·1**

1. $\int_{-10}^{10}(3x^2+4x-5)\,dx=(x^3+2x^2-5x)\Big|_{-10}^{10}=(10^3+2\cdot10^2-5\cdot10)-((-10)^3+2\cdot(-10)^2-5\cdot(-10))=1900$

2. $\int_{-50}^{30}8\,dx=(8x)\Big|_{-50}^{30}=(8\cdot30)-8\cdot(-50)=640$

3. $\int_2^7\frac{x^5}{x^2}\,dx=\int_2^7 x^3\,dx=\left(\frac{x^4}{4}\right)\Big|_2^7=\left(\frac{7^4}{4}\right)-\left(\frac{2^4}{4}\right)=\frac{2385}{4}=596.25$

4. $\int_6^{36}\frac{1}{t}\,dt=(\ln t)\Big|_6^{36}=(\ln 36)-(\ln 6)=\ln\left(\frac{36}{6}\right)=\ln 6$

5. $\int_{0.5\pi}^{\pi}\sec\left(\frac{5}{6}\theta\right)\tan\left(\frac{5}{6}\theta\right)d\theta=\frac{6}{5}\sec\left(\frac{5}{6}\theta\right)\Big|_{0.5\pi}^{\pi}=\left(\frac{6}{5}\sec\left(\frac{5}{6}\pi\right)\right)-\left(\frac{6}{5}\sec\left(\frac{5}{6}\cdot0.5\pi\right)\right)=$

$\frac{6}{5}\left(\sec\left(\frac{5\pi}{6}\right)-\sec\left(\frac{5\pi}{12}\right)\right)\approx-6.0221$

6. $\int_1^{\sqrt3}\frac{dx}{\sqrt{4-x^2}}=\int_1^{\sqrt3}\frac{dx}{\sqrt{2^2-x^2}}=\sin^{-1}\left(\frac{x}{2}\right)\Big|_1^{\sqrt3}=\sin^{-1}\left(\frac{\sqrt3}{2}\right)-\sin^{-1}\left(\frac{1}{2}\right)=\frac{\pi}{3}-\frac{\pi}{6}=\frac{\pi}{6}\approx0.5236$

7. $\int_1^2(3x^4-5x^3-21x^2+36x-10)\,dx=\left(\frac{3x^5}{5}-\frac{5x^4}{4}-7x^3+18x^2-10x\right)\Big|_1^2=$

$\left(\frac{3\cdot2^5}{5}-\frac{5\cdot2^4}{4}-7\cdot2^3+18\cdot2^2-10\cdot2\right)-\left(\frac{3\cdot1^5}{5}-\frac{5\cdot1^4}{4}-7\cdot1^3+18\cdot1^2-10\cdot1\right)=-\frac{103}{20}=-5.15$

8. $\int_3^5(x^3\ln x)\,dx=\left(\frac{x^4\ln x}{4}-\frac{x^4}{16}\right)\Big|_3^5=\left(\frac{5^4\ln 5}{4}-\frac{5^4}{16}\right)-\left(\frac{3^4\ln 3}{4}-\frac{3^4}{16}\right)\approx195.2278$

9. $\int_1^{\sqrt3}\cot^{-1}(x)\,dx=\left(x\cot^{-1}x+\frac{1}{2}\ln|1+x^2|\right)\Big|_1^{\sqrt3}=\left(\sqrt3\cot^{-1}\sqrt3+\frac{1}{2}\ln|1+(\sqrt3)^2|\right)-$

$\left(1\cot^{-1}(1)+\frac{1}{2}\ln|1+(1)^2|\right)=\left(\sqrt3\cot^{-1}\sqrt3+\frac{1}{2}\ln(4)\right)-\left(\cot^{-1}(1)+\frac{1}{2}\ln(2)\right)=\left(\sqrt3\cdot\frac{\pi}{6}+\frac{1}{2}\ln(4)\right)-$

$\left(\frac{\pi}{4}+\frac{1}{2}\ln(2)\right)\approx0.4681$

10. $\int_2^5\frac{1}{1+e^x}\,dx=(x-\ln(1+e^x))\Big|_2^5=(5-\ln(1+e^5))-(2-\ln(1+e^2))=3-\ln(1+e^5)+\ln(1+e^2)\approx0.1202$

**9·2**

1. By Property 1, $\int_2^2 f(x)\,dx=0$.
2. By Property 2, $\int_0^{-2} f(x)\,dx=-\int_{-2}^0 f(x)\,dx=-12$.
3. By Property 1, $\int_1^1 f(x)\,dx=0$.
4. By Property 3, $\int_{-2}^2 f(x)\,dx=\int_{-2}^0 f(x)\,dx+\int_0^2 f(x)\,dx=12+15=27$.
5. By Property 4, $\int_{-2}^0 5f(x)\,dx=5\int_{-2}^0 f(x)\,dx=5\cdot12=60$.
6. By Properties 4 and 3, $\int_2^{-2}10f(x)\,dx=10\int_2^{-2}f(x)\,dx=-10\left[\int_{-2}^0 f(x)\,dx+\int_0^2 f(x)\,dx\right]=$

   $-10\cdot(12+15)=-10\cdot(27)=-270$.

7. By Property 5, $\int_1^5 [f(x)+g(x)]dx = \int_1^5 f(x)dx + \int_1^5 g(x)dx = -8+22 = 14.$

8. By Property 5, $\int_1^5 [f(x)-g(x)]dx = \int_1^5 f(x)dx - \int_1^5 g(x)dx = -8-22 = -30.$

9. By Property 4, $\int_1^5 \frac{1}{2}f(x)dx = \frac{1}{2}\int_1^5 f(x)dx = \frac{1}{2}(-8) = -4.$

10. By Property 4, $\int_1^5 2g(x)dx + \int_1^5 3f(x)dx = 2\int_1^5 g(x)dx + 3\int_1^5 f(x)dx = 2(22)+3(-8) = 20.$

**9·3**

1. $\frac{d}{dx}\left[\int_0^x (t^2+3)^{-5}dt\right] = (x^2+3)^{-5} = \frac{1}{(x^2+3)^5}$

2. $\frac{d}{dx}\left[\int_1^x \sqrt{3t+5}\,dt\right] = \sqrt{3x+5}$

3. $\frac{d}{dx}\left[\int_\pi^{x^4} t\sin t\,dt\right] = (x^4)\sin(x^4)(4x^3) = 4x^7\sin(x^4)$

4. $\frac{d}{dx}\left[\int_{-5}^{5x^2} \sqrt[3]{t^2}\,dt\right] = \sqrt[3]{(5x^2)^2}(10x) = 10x\sqrt[3]{25x^4} = 10x^2\sqrt[3]{25x}$

5. $\frac{d}{dx}\left[\int_{-10}^{x+2} (t^2-2t+1)dt\right] = (x+2)^2 - 2(x+2)+1 = x^2+2x+1$

6. If $F(x) = \int_0^x \sin(3t)dt$, then $F'(x) = \sin(3x)$.

7. If $F(x) = \int_5^{4x} \frac{1}{t+1}dt$, then $F'(x) = \frac{1}{(4x)+1}\cdot 4 = \frac{4}{4x+1}$.

8. If $F(x) = \int_0^{\sin x} 6t^2dt$, then $F'(x) = 6(\sin x)^2\cos x = 6\sin^2 x\cos x$.

9. If $F(x) = \int_{-3}^{\sqrt{x}} 2t^4dt$, then $F'(x) = 2(\sqrt{x})^4\cdot\frac{1}{2}x^{-\frac{1}{2}} = x^{\frac{3}{2}}$.

10. If $F(x) = \int_{-8}^{2x+1} (3t-7)dt$, then $F'(x) = (3(2x+1)-7)\cdot 2 = 12x-8$.

**9·4**

Note: In these exercises, the symbol $\Rightarrow$ means "implies."

1. By the Mean Value Theorem, you have $\int_{-1}^1 (2x+6)dx = (2c+6)(1-(-1)) \Rightarrow (x^2+6x)\Big|_{-1}^1 = (2c+6)(2) \Rightarrow$ $((1)^2+6(1))-((-1)^2+6(-1)) = 4c+12 \Rightarrow 12 = 4c+12 \Rightarrow 0 = c.$

2. By the Mean Value Theorem, you have $\int_0^4 (2-5\sqrt{x})dx = (2-5\sqrt{c})(4-0) \Rightarrow \left(2x - \frac{10}{3}x^{\frac{3}{2}}\right)\Big|_0^4 =$ $(2-5\sqrt{c})(4) \Rightarrow \left(2(4) - \frac{10}{3}\cdot 4^{\frac{3}{2}}\right) - \left(2(0) - \frac{10}{3}\cdot 0^{\frac{3}{2}}\right) = 8-20\sqrt{c} \Rightarrow 8 - \frac{80}{3} = 8-20\sqrt{c} \Rightarrow \frac{16}{9} = c.$

3. By the Mean Value Theorem, you have $\int_1^4 \frac{4}{x^3}dx = \left(\frac{4}{c^3}\right)(4-1) \Rightarrow \left(-\frac{2}{x^2}\right)\Big|_1^4 = \left(\frac{4}{c^3}\right)(3) \Rightarrow$ $\left(-\frac{2}{4^2}\right) - \left(-\frac{2}{1^2}\right) = \frac{12}{c^3} \Rightarrow \frac{15}{8} = \frac{12}{c^3} \Rightarrow 2\sqrt[3]{\frac{4}{5}} = c.$

4. By the Mean Value Theorem, you have $\int_0^\pi \sin x\,dx = (\sin c)(\pi - 0) \Rightarrow (-\cos x)\Big|_0^\pi = (\sin c)(\pi) \Rightarrow$ $(-\cos\pi)-(-\cos 0) = \pi\sin c = (\sin c)(\pi) \Rightarrow (-(-1))-(-(1)) = \pi\sin c \Rightarrow 2 = \pi\sin c \Rightarrow \frac{2}{\pi} =$ $\sin c \Rightarrow \sin^{-1}\left(\frac{2}{\pi}\right) = c.$

5. By the Mean Value Theorem, you have $\int_1^3 \frac{1}{x}dx = \left(\frac{1}{c}\right)(3-1) \Rightarrow (\ln|x|)\Big|_1^3 = \left(\frac{1}{c}\right)(2) \Rightarrow \ln 3 - \ln 1 =$

$\left(\frac{1}{c}\right)(2) \Rightarrow \ln 3 - 0 = \frac{2}{c} \Rightarrow \ln 3 = \frac{2}{c} \Rightarrow \frac{2}{\ln 3} = c.$

6. $\frac{1}{b-a}\int_a^b f(x)dx = \frac{1}{(2-(-2))}\int_{-2}^{2} x^2\,dx = \frac{1}{4}\left(\frac{x^3}{3}\right)\Bigg|_{-2}^{2} = \frac{1}{4}\left(\frac{2^3}{3} - \frac{(-2)^3}{3}\right) = \frac{1}{4}\left(\frac{8}{3}+\frac{8}{3}\right) = \frac{4}{3}.$

7. $\frac{1}{b-a}\int_a^b f(x)dx = \frac{1}{(3-1)}\int_1^3 \frac{1}{x}dx = \frac{1}{2}(\ln x)\Big|_1^3 = \frac{1}{2}(\ln 3 - \ln 1) = \frac{1}{2}(\ln 3 - 0) = \frac{\ln 3}{2}.$

8. $\frac{1}{b-a}\int_a^b f(x)dx = \frac{1}{\left(\frac{\pi}{2} - \left(-\frac{\pi}{2}\right)\right)}\int_{-\frac{\pi}{2}}^{\frac{\pi}{2}} \cos x\,dx = \frac{1}{\pi}(\sin x)\Big|_{-\frac{\pi}{2}}^{\frac{\pi}{2}} = \frac{1}{\pi}\left(\sin\left(\frac{\pi}{2}\right) - \sin\left(-\frac{\pi}{2}\right)\right) = \frac{1}{\pi}(1-(-1)) = \frac{2}{\pi}.$

9. $\frac{1}{b-a}\int_a^b f(x)dx = \frac{1}{(4-1)}\int_1^4 \frac{9}{2}\sqrt{x}\,dx = \frac{1}{3}\left(3x^{\frac{3}{2}}\right)\Big|_1^4 = \left(4^{\frac{3}{2}} - 1^{\frac{3}{2}}\right) = (8-1) = 7.$

10. $\frac{1}{b-a}\int_a^b f(x)dx = \frac{1}{(1-0)}\int_0^1 e^x\,dx = \frac{1}{1}(e^x)\Big|_0^1 = e^1 - e^0 = e - 1.$

# APPLICATIONS OF THE DERIVATIVE AND THE DEFINITE INTEGRAL

## 10 Applications of the derivative

**10·1**

1. $f'(x) = 3x^2 + e^x + \cos(x)$ and thus $f'(-1) = 3 + \frac{1}{e} + \cos(-1) = 3 + \frac{1}{e} + \cos(1).$
2. $f'(x) = \frac{1}{x-1} + 2x$ and thus $f'(2) = 5.$
3. $y' = 4x + 4$ hence $y'(-2) = -4$ so the required equation is $y = -4(x+2) = -4x - 8.$
4. $f'(x) = 3x^2 - 12x + 9.$
5. $f'(x) = 2x - \frac{2}{\sqrt{x}}$. Set $f'(x) = 0$ and solve to get $\frac{2x^{\frac{3}{2}} - 2}{\sqrt{x}} = 0$ or $x = 1$. Thus the only point is (1, –2).
6. Setting $f'(x) = 0$ and solving you get $f'(x) = 5x^4 - 15x^2 - 20 = 5(x^4 - 3x^2 - 4) =$ $5(x^2 - 4)(x^2 + 1) = 0$ whose solution is $x = \pm 2$ so the points are (2, –41) and (–2, 55).
7. Differentiating implicitly you get $2x + 3xy' + 3y + 2yy' = 0$, so $y' = -\frac{2x+3y}{3x+2y}$ and at (1, 1) you get the slope $y' = -1$. This gives the required equation to be $y - 1 = -1(x-1)$ and simplifying you get $y = -x + 2.$
8. $y' = 4x$ and to be parallel to $y = 8x + 3$ the slopes must be equal, so $x$ must be 2. Also the tangent point must be on the curve and the point is therefore (2, 11). The required equation is then $y - 11 = 8(x-2)$ or simply $y = 8x - 5.$
9. There is no solution since (1, 2) is not on the curve.
10. $f'(x) = \frac{(x+1)(-\cos x) - (1 - \sin x)}{(x+1)^2}$ and so $f'(0) = -2.$ and the required equation is $y - 1 = -2(x-0)$ or simply $y = -2x + 1.$

**10·2**

1. $f'(t)=80-40t$ and $f'\left(\dfrac{3}{2}\right)=20$ acres per hour
2. $v'(t)=-32$ ft/sec²
3. $y'=3+16x-3x^2$ and at $t=2, y'=23$ castings per hour.
4. $s'(t)=200t+100$ and if $s(t)=39$ then solve $100t^2+100t=39$ to get $t=\dfrac{3}{10}$ or $-\dfrac{13}{10}$. Reject the negative value and the velocity is $s'\left(\dfrac{3}{10}\right)=160$ cm/sec.
5. $v(t)=s'(t)=4t^3-18t^2+24t-10=2(2t-5)(t-1)^2$ and $a(t)=v'(t)=s''(t)=12t^2-36t+24=12(t-1)(t-2)$. At $t=2.5$, $v=0$, but $a\neq 0$, so the direction changes but $v$ does not change sign at $t=1$. Both $v$ and $a$ are 0 at $t=1$, so no information is known.
6. $v(t)=s'(t)=\dfrac{3t^2}{2}-2$ and $a(t)=v'(t)=s''(t)=3t$. At $t=2$, $v=4$ ft/sec and $a=6$ ft/sec².
7. $y'=24-\dfrac{24x}{5}$ and $y'(3)=24\left(1-\dfrac{3}{5}\right)=\dfrac{48}{5}$ pints/lb.
8. $v(t)=s'(t)=-32t$ and $a(t)=v'(t)=-32$ so the velocity at 2 seconds is $-64$ ft/sec and the acceleration is $-32$ ft/sec².
9. $v(t)=112-32t$ and $a(t)=-32$ and when $t=3$, $v=16$ ft/sec. Also, $v=0$ when $t=3.5$ so the maximum height is $s(3.5)=196$ ft.
10. $V'(t)=250(-80+2t)$ so $V'(5)=-17{,}500$ gal/min.

**10·3**

1. Technically, there is a discontinuity at $x=3$, and thus the function is not differentiable there. However, the discontinuity is removable and on removal the resulting function is differentiable there.
2. $f'(0)=\lim\limits_{x\to 0}\dfrac{|x|-0}{x-0}=\lim\limits_{x\to 0}\dfrac{|x|}{x}$. But $\lim\limits_{x\to 0^+}\dfrac{|x|}{x}=\lim\limits_{x\to 0^+}\dfrac{x}{x}=1$ and $\lim\limits_{x\to 0^-}\dfrac{|x|}{x}=\lim\limits_{x\to 0^-}\dfrac{-x}{x}=-1$, so the limit does not exist and the function is not differentiable at $x=0$. On the other hand, $f'(x)=1$ when $x>0$, and $f'(x)=-1$ when $x<0$.
3. $\lim\limits_{x\to 2}(x-2)^{\frac{1}{3}}=0$ and thus the function is continuous there but $\lim\limits_{x\to 2}\dfrac{(x-2)^{\frac{1}{3}}-0}{x-2}=\lim\limits_{x\to 2}\dfrac{1}{(x-2)^{\frac{2}{3}}}$ does not exist and the function is not differentiable at $x=2$. In fact, there is a vertical tangent at (2, 0).
4. $\lim\limits_{h\to 0^+}\dfrac{f(3+h)-f(3)}{h}=\lim\limits_{h\to 0^+}\dfrac{(-4-(3+h)^2)-(-13)}{h}=\lim\limits_{h\to 0^+}\dfrac{-6h-h^2}{h}=\lim\limits_{h\to 0^+}(-6-h)=-6$ and

   $\lim\limits_{h\to 0^-}\dfrac{f(3+h)-f(3)}{h}=\lim\limits_{h\to 0^-}\dfrac{(5-6(3+h)-(-13)}{h}=\lim\limits_{h\to 0^-}\dfrac{-6h}{h}=-6$ so the derivative exists at $x=3$.
5. $\lim\limits_{x\to 0^+}\dfrac{f(x)-f(0)}{x-0}=\lim\limits_{x\to 0^+}\dfrac{x^2}{x}=\lim\limits_{x\to 0^+}x=0$ and $\lim\limits_{x\to 0^-}\dfrac{f(x)-f(0)}{x-0}=\lim\limits_{x\to 0^+}\dfrac{x-2}{x}$ and this limit does not exist so the function is not differentiable at $x=0$.

**10·4**

1. a. $f'(x)=x^3-3x^2+2x=x(x^2-3x+2)=x(x-1)(x-2)$ so the critical numbers are 0, 1, and 2 and the corresponding critical values are $0, \dfrac{1}{4}$, and 0.

   b. The derivative is positive on [0, 1] and [2, ∞) and negative on (−∞, 0] and [1, 2] and the function is increasing on [0, 1] and [2, ∞) and decreasing on (−∞,0] and [1, 2].

   c. $f(0)=0$ and $f(2)=0$ are relative minimums and $f(1)=\dfrac{1}{4}$ is a relative maximum.

2. a. $f'(x)=\frac{1}{2x^{\frac{1}{2}}}+\frac{1}{2x^{\frac{3}{2}}}=\frac{1}{2\sqrt{x}}\left(1+\frac{1}{x}\right)=\frac{x+1}{2x^{\frac{3}{2}}}$ and this is equal to 0 when $x=-1$, but that value is not in the domain so there are no critical points or critical values.

   b. The domain is $(0, \infty)$ and the function is increasing over its domain since the derivative is positive there.

   c. There are no relative extrema.

3. a. $f'(x)=2\cos\left(\frac{x}{2}\right)$ and this is equal to 0 when $\frac{x}{2}=\frac{(2n-1)\pi}{2}$ or when $x=(2n-1)\pi$ for $n$ an integer. The critical values are $\pm 4$ at these values.

   b. The function is increasing on $[(4k-1)\pi,(4k+1)]$ and decreasing on $[(4k+1)\pi,(4k+3)\pi]$ for integers $k$.

   c. The function has relative maximums of 4 at $x=(4k+1)\pi$ and relative minimums of $-4$ at $x=(4k+3)\pi$.

4. a. $f'(x)=\frac{x^2}{\sqrt{x^2-9}}+\sqrt{x^2-9}=\frac{2x^2-9}{\sqrt{x^2-9}}$, which is 0 when $x=\pm\frac{3}{\sqrt{2}}$ but the function is undefined at these points. There are no critical points or critical values.

   b. The derivative is positive on $(-\infty, -3)$ and $(3, \infty)$. The function is undefined on $(-3, 3)$.

   c. There are no relative extrema.

5. a. $\lim_{x\to -2^+}\frac{x^2+1-5}{x+2}=\lim_{x\to -2^+}\frac{(x-2)(x+2)}{x+2}=-4$, but $\lim_{x\to -2^-}\frac{2(x+2)}{x+2}=2$, so the derivative does not exist at $x=-2$. Also, $f'(0)=0$ so the critical points are $-2$ and 0. The corresponding critical values are $f(-2)=5$ and $f(0)=1$.

   b. The derivative is positive on $(-\infty,-2)$ and $(0,\infty)$ and negative on $(-2, 0)$.

   c. There is a relative maximum value of $f(-2)=5$ and a relative minimum value of $f(0)=1$.

6. a. $f'(x)=\frac{-2}{(x-4)^{\frac{1}{3}}}$ and is thus undefined at $x=4$, which is the only critical point and the corresponding critical value is $f(4)=2$.

   b. The derivative is positive on $(-\infty, 4)$ and negative on $(4, \infty)$.

   c. There is a relative maximum of 2 at $x=4$.

7. $f'(x)=3x^2+2ax$ and for a relative extreme to occur at $(2, 3)$, $f(2)=8+4a+b=3$ and since $f'(2)$ must be 0, you get $12+4a=0$ or $a=-3$ and thus $b=7$.

8. The constraints are the geometric figures and the amount of wire. So $0\le x\le 25$ and $0\le s\le 20$ where $x$ and $s$ denote length of the sides of the square and pentagon, respectively. Also, $4x+5s=100$ from which it follows that $s=\frac{100-4x}{5}$. For ease, let $a=\frac{5\cot(36^\circ)}{2}$. Then the total area is given by $T=x^2+as^2$ and $T'=2x+2ass'$ by implicit differentiation. Setting this equal to 0 and solving for $x$ you get $2x-\frac{8}{5}a\left(\frac{100-4x}{5}\right)=0$ and simplifying you get $x\left(2+\frac{32}{25}a\right)=\frac{800a}{25}$ or $x=\frac{800a}{50+32a}=\frac{400a}{25+16a}$ but $T''>0$ at this value indicating a minimum, which is not what you wanted. So the maximum value is achieved at $x=5$ which is an interval endpoint. Thus put all the wire on the square.

9. Since the amount $A$ is proportional to the square of the interest rate $r$, assume $A=kr^2$. Then the profit, $P=.09A-rA=k(.09r^2-r^3)$ and $P'=.18kr-3kr^2=3kr(.06-r)$ and this is 0 when $r=.06$ and this will be a maximum since $P''$ is negative at this point.

10. Let $C$ denote the total cost, $h$ the height of the jar, $r$ the radius of the jar top, $a$ the cost of the glass per unit area and $V$ the volume. Then $C = a\pi r^2 + a(2\pi rh) + 3a(\pi r^2) = 4a\pi r^2 + 2a\pi rh$ and $V = \pi r^2 h$. Now $\frac{dC}{dr} = 8\pi ar + 2\pi a\left(r\frac{dh}{dr} + h\right)$ and from the equation for $V$ you get $0 = \pi\left(r^2\frac{dh}{dr} + 2rh\right)$ from which you get $\frac{dh}{dr} = -\frac{2h}{r}$. Substitute this into the equation above to get $\frac{dC}{dr} = 8\pi ar - 4\pi ah + 2\pi ah = 8\pi ar - 2\pi ah$. Now set $\frac{dC}{dr} = 0$ to get $h = 4r$. Finally, $\frac{d^2C}{dr^2} = 8\pi a - 2\pi a\frac{dh}{dr} = 8\pi a + 16\pi a > 0$. So $h = 4r$ give a minimum.

**10·5**

1. $f'(x) = 4x^3 - 6$ and $f''(x) = 12x^2$, so 0 is a possible point of inflection, but since the second derivative is positive for all non-zero values, the curve is concave up and (0, 2) is not a point of inflection.
2. $f'(x) = 4x^3 - 18x^2 + 24x - 8$ and $f''(x) = 12x^2 - 36x + 24$ and $f'''(x) = 24x - 36$. The possible points of inflection are determined by $x = 1, 2$ and since $f'''(x) \neq 0$ at either of these points, both determine points of inflection. Employ the little-used fact that if $f''(c) = 0$ and $f'''(c) \neq 0$, then $(c, f(c))$ is a point of inflection. The points of inflection are (1, –1) and (2, 0)
3. This is a straight line graph and has no points of inflection.
4. $y' = 3 + \frac{3}{5(x+2)^{\frac{2}{5}}}$ and this is undefined for $x = -2$. Also $y'' = \frac{-6}{25(x+2)^{\frac{7}{5}}}$ and $y'' > 0$ (concave up) when $x < -2$, and $y'' < 0$ (concave down) when $x > -2$, so (–2, –6) is a point of inflection.
5. $f'(x) = \begin{cases} 2x & \text{if } x < 2 \\ -2x & \text{if } x > 2 \end{cases}$ and $f''(x) = \begin{cases} 2 & \text{if } x < 2 \\ -2 & \text{if } x > 2 \end{cases}$. The function is not continuous at 2 and thus has no derivative there and also no tangent. The graph changes concavity at (2, 3), but it is not a point of inflection.
6. $f'(x) = \begin{cases} 3x^2 & \text{if } x < 0 \\ 4x^3 & \text{if } x \geq 0 \end{cases}$ and $f''(x) = \begin{cases} 6x & \text{if } x < 0 \\ 12x^2 & \text{if } x \geq 0 \end{cases}$. Solve $f''(x) = 0$ to get $x = 0$. The graph is concave up for $x > 0$ and concave down for $x < 0$, so (0, 0) is a point of inflection.
7. $f'(x) = 6\cos(3x)$ and $f''(x) = -18\sin(3x)$. Solve $f''(x) = 0$ to get $x = 0$ to possibly determine a point of inflection. The curve is concave up for $x < 0$ and concave down for $x > 0$ and (0, 0) is a point of inflection.
8. $y' = \frac{1}{3(x-1)^{\frac{2}{3}}}$ and $y'' = \frac{-2}{9(x-1)^{\frac{5}{3}}}$ The function is continuous at $x = 1$, but $y' = \frac{1}{3(x-1)^{\frac{2}{3}}}$ and $y'' = \frac{-2}{9(x-1)^{\frac{5}{3}}}$ do not exist at $x = 1$. However, since $y'$ approaches $\infty$ as $x$ approaches 1, there is a vertical tangent at (1, –2). Also, $y''$ is positive when $x < 1$ and negative when $x > 1$, so the concavity changes at $x = 1$. Thus, (1, –2) is a point of inflection.
9. $y' = 4x^3 - 36x$ and $y'' = 12x^2 - 36 = 12(x - \sqrt{3})(x + \sqrt{3})$. Thus, $\pm\sqrt{3}$ possibly determine points of inflection. In fact the curve is concave up for $x < -\sqrt{3}$ or $x > \sqrt{3}$ and concave down for $-\sqrt{3} < x < \sqrt{3}$, so the points of inflection are $(\pm\sqrt{3}, -44)$.
10. $h'(x) = 3ax^2 + 2bx + c$ and $h''(x) = 6ax + 2b$. Since a relative max will occur at (0, 3), $h'(0) = 0$ so $c = 0$. Also, $h(0) = 3$ implies that $d = 3$. $h''(1) = 0$ implies that $6a + 2b = 0$ and $h(1) = -1$ implies that $a + b + c + d = -1$. So you get $3a + b = 0$ and $a + b = -4$. Solving this set of equations gives $a = 2$ and $b = -6$. The solution equation is $y = 2x^3 - 6x^2 + 3$.

**10·6**

1. $f'(x) = 3x^2 + 1$, so solve $3c^2 + 1 = \frac{f(2) - f(-1)}{2 - (-1)} = \frac{12}{3} = 4$ to get $c = 1$ and $c = -1$.
2. There is no solution since the function is not continuous in the interval, particularly at 0.
3. $g'(x) = 1 - \frac{1}{2\sqrt{x}}$, so solve $1 - \frac{1}{2\sqrt{c}} = \frac{2}{3}$ or $\frac{1}{2\sqrt{c}} = \frac{1}{3}$ or $2\sqrt{c} = 3$ or finally $c = \frac{9}{4}$.

4. There is no solution since the function is discontinuous at 1.

5. Solve $3c^2 - 6c = 1$ to get $c = \frac{6 \pm \sqrt{36+12}}{6} = \frac{6 \pm 4\sqrt{3}}{6} = \frac{3 \pm 2\sqrt{3}}{3}$.

6. There is no solution since the derivative does not exist at 0.

7. Solve $1 + \sin(c) = \frac{\left(\frac{3\pi}{2} - \cos\left(\frac{3\pi}{2}\right)\right) - (\pi - \cos(\pi))}{\frac{3\pi}{2} - \pi} = \frac{\frac{\pi}{2} - 1}{\frac{\pi}{2}} = 1 - \frac{2}{\pi}$ or $\sin(c) = -\frac{\pi}{2}$, which solves as $c = \arcsin\left(-\frac{\pi}{2}\right)$.

8. Solve $24c^2 + 36c + 3 = \frac{22 - (-5)}{3} = \frac{27}{3} = 9$ to obtain $c = \frac{-3 \pm \sqrt{13}}{4}$.

9. $f'(x) = \frac{(1+\cos x)(\cos x) - \sin x(-\sin x)}{(1+\cos x)^2} = \frac{\cos x + \cos^2 x + \sin^2 x}{(1+\cos x)^2} = \frac{1+\cos x}{(1+\cos x)^2}$; so solve $\frac{1}{(1+\cos c)} = \frac{1}{\frac{\pi}{2}}$ or $1 + \cos c = \frac{\pi}{2}$, which solves as $c = \arccos\left(\frac{\pi}{2} - 1\right)$.

10. Let $f(x) = \ln x$ and consider the interval $\left[1, \frac{8}{7}\right]$. By the MVT you get $\frac{1}{c} = \frac{\ln\left(\frac{8}{7}\right) - \ln(1)}{\frac{8}{7} - 1}$ or on simplification $\frac{1}{c} = 7\ln\left(\frac{8}{7}\right)$ or finally that $c = \frac{1}{7\ln\left(\frac{8}{7}\right)}$. Also you know that $1 < c < \frac{8}{7}$. So you get $1 < \frac{1}{7\ln\left(\frac{8}{7}\right)} < \frac{8}{7}$ and solving this inequality you get $\frac{1}{8} < \ln\left(\frac{8}{7}\right) < \frac{1}{7}$.

# 11 Applications of the definite integral

**11·1**

1. $\int_3^6 (2x^2 + 2x - 24)\,dx = \left[\frac{2x^3}{3} + x^2 - 24x\right]_3^6 = \left[\left(\frac{2(216)}{3} + 36 - 24(6)\right) - \left(\frac{2(27)}{3} + 9 - 24(3)\right)\right] =$ $[144 + 36 - 144 - 18 - 9 + 72] = 81$

2. $\int_{\frac{\pi}{3}}^{\frac{2\pi}{3}} \sin x\,dx = [-\cos x]_{\frac{\pi}{3}}^{\frac{2\pi}{3}} = \left[-\cos\left(\frac{2\pi}{3}\right) + \cos\left(\frac{\pi}{3}\right)\right] = \frac{1}{2} + \frac{1}{2} = 1$

3. $\int_1^3 (8x - 2x^2)\,dx = \left[4x^2 - \frac{2x^3}{3}\right]_1^3 = \frac{44}{3}$

4. $\int_0^{\frac{\pi}{4}} \sec^2 x\,dx = [\tan x]_0^{\frac{\pi}{4}} = 1$

5. $\int_0^8 \sqrt{4x+4}\,dx = 2\int_0^8 \sqrt{x+1}\,dx = \left[\frac{2(x+1)^{\frac{3}{2}}}{\frac{3}{2}}\right]_0^8 = \left[\frac{4(9)^{\frac{3}{2}}}{3} - \frac{4}{3}\right] = \frac{4(27) - 4}{3} = \frac{104}{3}$

6. $\int_0^{\frac{\pi}{6}} \cos x\,dx = [\sin x]_0^{\frac{\pi}{6}} = \frac{1}{2}$

**11·2**

1. $y = 4 - x^2$ intersects the $x$-axis at –2 and 2. Thus, the desired area $= \int_{-2}^{2}(4 - x^2)dx = \left[4x - \frac{x^3}{3}\right]_{-2}^{2} = \left(8 - \frac{8}{3}\right) - \left(-8 + \frac{8}{3}\right) = 16 - \frac{16}{3} = \frac{32}{3}$.

2. Solve for the intersection points by setting $x^2 = x + 2$, which has solutions $x = -1$ and 2. Thus the area is $\int_{-1}^{2}((x+2) - x^2)dx = \left[\frac{x^2}{2} + 2x - \frac{x^3}{3}\right]_{-1}^{2} = \left(6 - \frac{8}{3}\right) - \left(-2 + \frac{5}{6}\right) = \frac{9}{2}$.

3. The curves intersect at $x = 0$ and 1. The area is $\int_{0}^{1}(\sqrt{x} - x^2)dx = \left[\frac{2x^{\frac{3}{2}}}{3} - \frac{x^3}{3}\right]_{0}^{1} = \frac{1}{3}$.

4. Solve for the intersection points by setting $(x+1)^3 = (x+1)$ to get $x = 0, -1$, and $-2$. Now, $(x+1)^3 - (x+1) = (x+1)((x+1)^2 - 1) = (x+1)(x^2 + 2x)$; which is less than or equal to 0 when $-1 \le x \le 0$ and is greater than or equal to 0 when $-2 \le x \le -1$, so the area is given by the following two integrals: $\int_{-2}^{-1}((x+1)^3 - (x+1))dx + \int_{-1}^{0}((x+1) - (x+1)^3)dx = \left[\frac{(x+1)^4}{4} - \frac{(x+1)^2}{2}\right]_{-2}^{-1} + \left[\frac{(x+1)^2}{2} - \frac{(x+1)^4}{4}\right]_{-1}^{0} = \frac{1}{2}$.

5. The area is given by two integrals since $y = 0$ dominates when $x < 0$ and is dominated when $x > 0$. The area is $\int_{-1}^{0}(-x^3 - x)dx + \int_{0}^{1}(x^3 + x)dx = \left[\frac{-x^4}{4} - \frac{x^2}{2}\right]_{-1}^{0} + \left[\frac{x^4}{4} + \frac{x^2}{2}\right]_{0}^{1} = \frac{3}{2}$.

6. The curves intersect when $2x + 3 = -x + 6$; that is, when $x = 1$. To find which function is greater, solve $2x + 3 \le -x + 6$ to find that this occurs when $x \le 1$. Thus, the area is $\int_{0}^{1}((-x+6) - (2x+3))dx = \int_{0}^{1}(-3x + 3)\,dx = \left[\frac{-3x^2}{2} + 3x\right]_{0}^{1} = \frac{3}{2}$.

7. The curves intersect when $e^x = e$; that is, when $x = 1$. When $0 \le x \le 1$, the curve $y = e$ is greater than $y = e^x$, so the area is $\int_{0}^{1}(e - e^x)dx = [ex - e^x]_0^1 = 1$.

8. Factoring the first equation you get $y = (x^4 - 2x^3 - x^2 + 2x) + 1 = x(x-1)(x+1)(x-2) + 1$. The curves intersect when their equations are equal; that is, when $x = 0$, 1, $-1$, and 2. To get an idea of the dominance in the region of interest notice that when $x > 2$, the quartic function is positive. From this observation, you can deduce the area is given by the integral $\int_{-1}^{0}(1 - x^4 + 2x^3 + x^2 - 2x - 1)dx = \left[x - \frac{x^5}{5} + \frac{2x^4}{4} + \frac{x^3}{3} - \frac{2x^2}{2} - x\right]_{-1}^{0} = -\frac{1}{5} - \frac{1}{2} + \frac{1}{3} + 1 = \frac{19}{30}$.

9. In the region of interest, the curves intersect when $3 - x^2 = -x + 1$; that is, when $x^2 - x - 2 = 0$, which occurs when $x = -1$ and $x = 2$. Reject 2 because it lies outside [–1, 1]. Also, the quadratic function dominates on [–1, 1], so the area is $\int_{-1}^{1}((3 - x^2) - (-x + 1))dx = \left[-\frac{x^3}{3} + \frac{x^2}{2} + 2x\right]_{-1}^{1} = \frac{10}{3}$.

10. The curves intersect when $x = \pm 1$ and $y = 1$ dominates between these values so the area is $\int_{-1}^{1}(1 - x^2)dx = \left[x - \frac{x^3}{3}\right]_{-1}^{1} = \frac{4}{3}$.

**11·3** In the following worked exercises, techniques of integration and integral tables are used. This approach is intended to illustrate and strengthen your use of these strategies. $L$ will be used to indicate the arc length in all problems.

1. The arc equation is $y=\frac{x^2}{2}; y'=x$ and then

$$L=\int_{-\sqrt{3}}^{0}\sqrt{1+x^2}\,dx=\left[\frac{x\sqrt{1+x^2}}{2}+\frac{\ln(x+\sqrt{1+x^2})}{2}\right]_{-\sqrt{3}}^{0}=\sqrt{3}-\frac{\ln(2-\sqrt{3})}{2} \text{ (Formula 61).}$$

2. The arc equation is $y=4-\frac{4x}{9}; y'=-\frac{4}{9}$ and then

$$L=\int_0^9\sqrt{1+\frac{16}{81}}\,dx=\frac{\sqrt{97}}{9}\int_0^9 dx=\left[\frac{\sqrt{97}}{9}x\right]_0^9=\sqrt{97}.$$

3. The arc equation is $y=\frac{(x^2+2)^{\frac{3}{2}}}{3}; y'=x(x^2+2)^{\frac{1}{2}}$ and then $L=\int_0^3\sqrt{1+x^2(x^2+2)}\,dx=\int_0^3\sqrt{(x^2+1)^2}\,dx=$

$$\int_0^3(x^2+1)dx=\left[\frac{x^3}{3}+x\right]_0^3=12.$$

4. The arc equation is $x=\frac{y^4+3}{6y}=\frac{y^3}{6}+\frac{1}{2y}$ and $x'=\frac{y^2}{2}-\frac{1}{2y^2}=\frac{1}{2}\left(\frac{y^4-1}{y^2}\right)$ and then $L=\int_1^2\sqrt{1+\frac{(y^4-1)^2}{4y^4}}\,dy=$

$$\int_1^2\sqrt{\frac{(y^4+1)^2}{4y^4}}\,dy=\frac{1}{2}\int_1^2\frac{(y^4+1)}{y^2}dy=\frac{1}{2}\int_1^2(y^2+y^{-2})dy=\frac{1}{2}\left[\frac{y^3}{3}-\frac{1}{y}\right]_1^2=\frac{1}{2}\left[\left(\frac{8}{3}-\frac{1}{2}\right)-\left(\frac{1}{3}-1\right)\right]=\frac{17}{12}.$$

5. The arc equation is $y=\frac{x^4}{4}+\frac{1}{8x^2}; y'=x^3-\frac{1}{4x^3}=\frac{4x^6-1}{4x^3}$ and then $L=\int_1^2\sqrt{1+\frac{(4x^6-1)^2}{16x^6}}\,dx=$

$$\int_1^2\sqrt{\frac{(4x^6+1)^2}{16x^6}}\,dx=\int_1^2\frac{(4x^6+1)}{4x^3}dx=\int_1^2\left(x^3+\frac{x^{-3}}{4}\right)dx=\left[\frac{x^4}{4}-\frac{1}{8x^2}\right]_1^2=\left(\frac{16}{4}-\frac{1}{32}\right)-\left(\frac{1}{4}-\frac{1}{8}\right)=\frac{123}{32}.$$

6. The arc equation is $y=\frac{\sqrt{x}(3x-1)}{3}; y'=\frac{3\sqrt{x}}{2}-\frac{1}{6\sqrt{x}}=\frac{18x-2}{12\sqrt{x}}=\frac{9x-1}{6\sqrt{x}}$ and then

$$L=\int_1^4\sqrt{1+\frac{(9x-1)^2}{36x}}\,dx=\int_1^4\sqrt{\frac{(9x+1)^2}{36x}}\,dx=\int_1^4\frac{9x+1}{6\sqrt{x}}dx=\int_1^4\left(\frac{3x^{\frac{1}{2}}}{2}+\frac{x^{-\frac{1}{2}}}{6}\right)dx=\left[x^{\frac{3}{2}}+\frac{x^{\frac{1}{2}}}{3}\right]_1^4=$$

$$\left[\left(8+\frac{2}{3}\right)-\left(1+\frac{1}{3}\right)\right]=\frac{22}{3}.$$

7. The arc equation is $y=\ln x; y'=\frac{1}{x}$ and then $L=\int_1^{\sqrt{3}}\sqrt{1+\frac{1}{x^2}}\,dx=\int_1^{\sqrt{3}}\frac{\sqrt{x^2+1}}{x}dx=$

$$\left[\sqrt{x^2+1}-\ln\left(\frac{1+\sqrt{x^2+1}}{x}\right)\right]_1^{\sqrt{3}}=\left[\left(2-\ln\left(\frac{3}{\sqrt{3}}\right)\right)-(\sqrt{2}-\ln(1+\sqrt{2})\right]=2-\sqrt{2}+\ln(1+\sqrt{2})-\frac{\ln 3}{2}.$$

(Formula 62 was used for the integration.)

8. The arc equation is $y=\frac{x^3}{3}+\frac{1}{4x}: y'=x^2-\frac{1}{4x^2}=\frac{4x^4-1}{4x^2}$ and then

$$L=\int_1^3\sqrt{1+\frac{(4x^4-1)^2}{16x^4}}\,dx=\int_1^3\sqrt{\frac{(4x^4+1)^2}{16x^4}}\,dx=\int_1^3\frac{4x^4+1}{4x^2}dx=\int_1^3\left(x^2+\frac{x^{-2}}{4}\right)dx=\left[\frac{x^3}{3}-\frac{1}{4x}\right]_1^3=$$

$$\left(\frac{27}{3}-\frac{1}{12}\right)-\left(\frac{1}{3}-\frac{1}{4}\right)=\frac{53}{6}.$$

9. The arc equation is $y^2 = \frac{x(x-3)^2}{9}$. Since $1 \le x \le 3, \sqrt{(x-3)^2} = 3-x$. Thus, the arc equation can be restated as $y = \frac{\sqrt{x}(3-x)}{3} = x^{\frac{1}{2}} - \frac{x^{\frac{3}{2}}}{3}; y' = \frac{x^{-\frac{1}{2}}}{2} - \frac{x^{\frac{1}{2}}}{2} = \frac{1}{2}\left(\frac{1}{\sqrt{x}} - \sqrt{x}\right) = \frac{1}{2}\left(\frac{1-x}{\sqrt{x}}\right)$ and then $L = \int_1^3 \sqrt{1 + \frac{(1-x)^2}{4x}}\, dx = \int_1^3 \sqrt{\frac{(x+1)^2}{4x}}\, dx = \int_1^3 \frac{x+1}{2\sqrt{x}} dx = \frac{1}{2}\int_1^3 \left(x^{\frac{1}{2}} + x^{-\frac{1}{2}}\right) dx = \frac{1}{2}\left[\frac{2x^{\frac{3}{2}}}{3} + 2x^{\frac{1}{2}}\right]_1^3 =$ $\frac{1}{2}\left[\left(\frac{2\sqrt{27}}{3} + 2\sqrt{3}\right) - \left(\frac{2}{3} + 2\right)\right] = 2\sqrt{3} - \frac{4}{3}$.

10. The arc equation is $y = 2(x-1)^{\frac{3}{2}}; y' = 3(x-1)^{\frac{1}{2}}$ and then $L = \int_1^{\frac{17}{9}} \sqrt{1 + 9(x-1)}\, dx = \int_1^{\frac{17}{9}} (9x-8)^{\frac{1}{2}}\, dx$. Now let $u = 9x - 8$, then $du = 9dx$; and when $x = 1$, $u = 1$ and when $x = \frac{17}{9}, u = 9$. Thus, you have $L = \frac{1}{9}\int_1^9 u^{\frac{1}{2}} du = \frac{1}{9}\left[\frac{2u^{\frac{3}{2}}}{3}\right]_1^9 = \frac{1}{9}\left(\frac{54}{3} - \frac{2}{3}\right) = \frac{52}{27}$.